VERY HIGH RESOLUTION SPECTROSCOPY

The Lord Rank, J.P., LL.D., 1888–1972

VERY HIGH RESOLUTION SPECTROSCOPY

edited by

R. A. SMITH
*Heriot-Watt University,
Edinburgh, Scotland*

1976

ACADEMIC PRESS
London · New York · San Francisco
A subsidiary of Harcourt Brace Jovanovich, Publishers

ACADEMIC PRESS INC. (LONDON) LTD.
24/28 Oval Road,
London NW1

United States Edition published by
ACADEMIC PRESS INC.
111 Fifth Avenue
New York, New York 10003

Copyright © 1976 by
RANK PRIZE FUNDS

All Rights Reserved
No part of this book may be reproduced in any form by photostat, microfilm, or any other means, without written permission from the publishers

Library of Congress Catalog Card Number: 75-19675
ISBN: 0-12-651650-2

Filmset by Ramsay Typesetting (Crawley) Ltd.
and printed in Great Britain by Whitstable Litho, Straker Brothers Ltd.

Participants at the Symposium

BERGEN, Mr. S. A., *Head of Research & Development, Pye Unicam Limited, York Street, Cambridge CB1 2PX, England*
BERLMAN, Professor I., *Professor of Physics, Microwave Division, The Hebrew University of Jerusalem, Jerusalem, Israel*
BIRKS, Dr. J. B., *Reader in Physics, University of Manchester, Atomic Molecular & Polymer Physics Group, Schuster Laboratory, The University, Manchester M13 9PL, England*
*†BRADLEY, Professor D. J., *Professor of Applied Optics, Imperial College of Science & Technology, London, SW7 2BZ, England*
BREWER, Dr. R. G., *IBM Fellow, IBM Research Laboratory, San José, California 95193, U.S.A.*
BROSSEL, Professor J., *Directeur, Laboratoire de Physique de l'Ecole Normale Superieure, 24, Rue Lhomond, 75231, Paris, France*
BULLOUGH, Professor R. K., *Professor of Mathematical Physics, University of Manchester Institute of Science and Technology, P.O. Box 88, Manchester M60 1QD, England*
*CHANCE, Professor Britton, *Chairman and Director, Johnson Research Foundation, Department of Biophysics and Physical Biochemistry, University of Pennsylvania, Philadelphia 19174, U.S.A.*
CLERC, Dr. M. C., *Research Scientist, French Atomic Energy Commission, Boite Postale No. 2, 91190 Gif-sur-Yvette, Saclay, France.*
COLLES, Dr. M. J., *Department of Physics, Heriot-Watt University, Riccarton, Currie, Edinburgh EH14 4AS, Scotland*
CONNERADE, Dr. J. P., *Lecturer, Imperial College of Science and Technology, Department of Physics, Prince Consort Road, London, S.W.7, England*
COOK, Professor A. H., F.R.S., *Jacksonian Professor of Natural Philosophy, University of Cambridge, Cavendish Laboratory, Free School Lane, Cambridge CB2 3RQ, England*
COURTENS, Dr. E. L., *Manager of Research Unit, International Business Machines Corpn, Research Laboratory Zurich, Saumerstrasse 4, CH-8803 Ruschlikon, Switzerland*
DAINTON, Professor Sir Frederick, F.R.S., *Chairman, University Grants Committee, 14, Park Crescent, London, W1N 4DH, England*
EISENTHAL, Dr. K. B., *Manager, Physical Sciences, IBM Research Laboratory, International Business Machines Corpn, Monterey & Cottle Roads, San Jose, California 95114, U.S.A.*
GIBSON, Professor A. F., *Head of Department of Physics, University of Essex, Wivenhoe*

LIST OF PARTICIPANTS

Park, Colchester CO4 3SQ, England
GILSON, Mr. A. R., Managing Director, Perkin-Elmer Limited, Beaconsfield, Bucks HP9 1QA, England
HAHN, Professor E. L., Professor of Physics, University of California, Berkeley, California 94720, U.S.A.
HANNA, Dr. D. C., Lecturer, University of Southampton, Department of Electronics, The University, Southampton SO9 5NH, England
*HARRIS, Professor S. E., Professor of Electrical Engineering, Microwave Laboratory, Stanford University, Stanford, California 94305, U.S.A.
HIGGINS, Dr. J. F., Director, Electro-Photonics Limited, 1, The Cutts, Dunmurry, Belfast BT17 9HN, N. Ireland.
HILL, Dr. R., F.R.S., Department of Biochemistry, University of Cambridge, Tennis Court Road, Cambridge, England
*JACQUINOT, Professor P., Member Académie des Sciences, Paris, CNRS., Laboratoire Aimé Cotton, Campus d'Orsay, Bâtiment 505, 91405 Orsay, France
JOHNSON, Professor F. A., Honorary Professor, Birmingham University, Deputy Director, Royal Radar Establishment, St. Andrews Road, Great Malvern, Worcs., WR14 3PS, England
*†JONES, Dr. F. E., M.B.E., F.R.S., Director, Philips Industries, 11, Hanover Square, London W1A 4QP, England
*KAISER, Professor W., Chairman, Physics Department, Technischen Universität München, 8 München 2, Arcisstrasse 21, West Germany
*KASTLER, Professor A., Professor of Physics, Université de Paris, Ecole Normale Supérieure, Laboratoire de Physique, 24 Rue Lhomond, Paris 5e, France
KEY, Dr. M. H., Senior Lecturer in Physics, Queen's University of Belfast, Belfast BT7 1NN, N. Ireland
KNIGHT, Dr. A. E. W., Research Fellow, Department of Chemistry, Indiana University, Chemistry Building, Bloomington, Indiana 47401, U.S.A.
LALLEMAND, Dr. P. M., Maître de Recherche, Centre National de la Recherche Scientifique, Laboratoire de Spectroscopic Hertzienne de l'Ecole Normale Superieure, 24, Rue Lhomond, 75231, Paris
LEHMANN, Professor J. C., Professor, Universite de Paris VI, Laboratoire de Spectroscopie Hertzienne de l'Ecole Normale Superieure, 11, Quai Saint Bernard, 75230, Paris
LOUDON, Professor R., Professor of Theoretical Physics, University of Essex, Wivenhoe Park, Colchester CO4 3SQ, England
†McGEE, Professor J. D., O.B.E. Sc.D., F.I.E.E., F.Inst.P., F.R.A.S., F.R.S., Emeritus Professor of Applied Physics, Imperial College of Science & Technology, London, England
†McLEAN, Dr. T. P., B.Sc., Ph.D., F.Inst.P., Head of Physics Department, Royal Radar Establishment, Great Malvern, Worcs., England
*MILLETT, Mr. E. J., Materials Group Leader, Solid State Physics Division, Mullard Research Laboratories, Redhill, Surrey, England
*MOORADIAN, Dr. A., Group Leader, Quantum Electronics Group, Massachusetts Institute of Technology, Lincoln Laboratory, Lexington, Mass. 02173, U.S.A.
MOORE, Dr. W. S., Senior Lecturer, University of Nottingham, Department of Physics, University Park, Nottingham NG7 2RD, England.
PIDGEON, Dr. C. R., Senior Lecturer, Department of Physics, Heriot-Watt University, Riccarton, Currie, Edinburgh EH14 4AS, Scotland
*PIKE, Dr. E. R., Deputy Chief Scientific Officer, Physics Group, Royal Radar Establishment, Great Malvern, Worcs. WR14 3PS, England

LIST OF PARTICIPANTS

PINARD, Dr. J., *Maître de Recherche, Centre National de la Recherche Scientifique, Laboratoire Aimé Cotton, Batiment 505, C.N.R.S. II Campus d'Orsay 91405 Orsay, France.*
PORTER, Professor Sir George, F.R.S., *Director and Fullerian Professor of Chemistry, The Royal Institution, 21, Albemarle Street, London, W1X 4BS, England*
PORTO, Professor S. P. S., *Instituto de Fisica, Universidade Estadual de Campinas, Caixa Postal 1170, 13100—Campinas, SP., Brazil*
*PRICE, Professor W. C., F.R.S., *Head of Physics Department, King's College, University of London, London, WC2R 2LS, England*
RAMSDEN, Mr. W., *Director and General Manager, Rank Hilger, Westwood, Margate, Kent CT9 4JL, England*
REID, Mr. E. S., *Research Assistant, Davy Faraday Research Laboratory, The Royal Institution, 21, Albemarle Street, London, W1X, England*
REYNOLDS, Professor C. T., *Professor of Physics, Princeton University, Princeton, N.J., U.S.A. (1973-1974—Churchill Overseas Fellow at Churchill College, Cambridge).*
RING, Professor J., *Professor of Physics, Department of Physics (Astronomy), Imperial College of Science & Technology, 10, Prince's Gardens, London, SW7 1NA, England*
SCHAEFER, Professor F. P., *Director, Max-Planck Institut fur Biophysikalische Chemie, Postfach 968, D-3400 Gottingen, Germany*
†SCHAGEN, Dr. P., O.B.E., Ph.D., F.Inst.P., *Head of Vacuum Physics Division, Mullard Research Laboratories, Redhill, Surrey, England*
SELINGER, Dr. B., *Senior Lecturer, Australian National University, Department of Chemistry, Box 4, Post Office, Canberra, A.C.T. 2600, Australia*
SERIES, Professor G. W., F.R.S., *Professor of Physics, University of Reading, J. J. Thomson Physical Laboratory, Whiteknights, Reading RG6 2AF, England*
SHIMODA, Professor K., *Professor of Physics, University of Tokyo, 3-1 Hongo 7-Chome, Bunkyo-Ku, Tokyo, Japan*
†SMITH, Dr. R. A., C.B.E., M.A., Ph.D., F.R.S.E., F.R.S., *Principal and Vice Chancellor, Heriot-Watt University, Edinburgh, Scotland*
SMITH, Dr. R. C., *Reader in Electronics, Department of Electronics, The University, Southampton SO9 5NH, England*
*SMITH, Professor S. D., *Head of Department of Physics, Heriot-Watt University, Riccarton, Midlothian, Scotland*
STEIN, Professor G., *Professor of Physical Chemistry, The Hebrew University of Jerusalem, P.O. Box 1255, Jerusalem 91000, Israel*
STILLWELL, Mr. P. F. T. C., *Director, Rank Research Laboratories Limited, Phoenix Works, Great West Road, Brentford, Middx, England*
SVELTO, Professor O., *Professor of Quantum Electronics, Instituto di Fisica del Politecnico, Piazza Leonardo da Vinci 32, 20133, Milano, Italy*
THOMAS, Dr. E. L., *Senior Lecturer, Department of Applied Physics, The University of Hull, Hull HU6 7RX, England*
*THOMPSON, Professor Sir Harold, C.B.E., F.R.S., *Professor of Chemistry, St. John's College, Oxford, England*
*TOWNES, Professor C. H., *Professor of Physics, University of California, Berkeley, California 94720, U.S.A.*
VON DER LINDE, Dr. D., *Member of Technical Staff, Bell Laboratories, 600, Mountain Avenue, Murray Hill, New Jersey 07974, U.S.A.*
VREHEN, Dr. Q. H. F., *Senior Research Scientist, Philips Research Laboratories, Eindhoven, Netherlands*
WALTHER, Professor H., *University Professor, University of Koln, Erstes Physikalisches*

LIST OF PARTICIPANTS

Institut, Universitastrasse 14, 5 Koln 41, Germany
WARE, Professor W. R., *Professor of Chemistry, The University of Western Ontario, London 72, Canada*
WEBB, Dr. C. E., *University Lecturer, Department of Physics, Clarendon Laboratory, Parks Road, Oxford, England*
WEBER, Professor G., *Professor of Biochemistry, University of Illinois, School of Chemical Sciences, Urbana 61801, U.S.A.*
WEST, Professor T. S., *Professor of Analytical Chemistry, Imperial College of Science and Technology, Department of Chemistry, Imperial Institute Road, London, SW7, England*
YAJIMA, Professor T., *Associate Professor, University of Tokyo, The Institute for Solid State Physics, Roppongi, Minato-Ku, Tokyo, Japan*

*Contributors to this volume.
†Members of the Advisory Committee on Opto-electronics.

Preface

The papers presented at The Rank Prize Funds' Symposium on "Very High Resolution Spectroscopy" have been collected together in this volume so as to be available to a larger group of scientists than those able to attend the Conference, which had to be rather strictly limited in numbers. The papers are arranged in the sequence in which they were delivered.

Some of the papers were available to the editor in written form and needed only small amendments to take account of points raised in the discussion which followed. Others were recorded and transcribed and, after editing, were sent to the authors for final revision. The figures were largely based on the authors' slides as presented at the Conference.

Following each paper a short discussion was held, and a longer period for discussion was available on the final afternoon of the Conference.

These discussions were recorded (with some difficulty) but have not all been reproduced in detail in the Conference Report. Authors were asked to "integrate" points made in these discussions in the final versions of their papers whenever possible.

It was naturally not found possible to deal with all the discussion in this way. A combined account of some of the main points of substance raised in the discussions after the papers, and of the general discussion on the final afternoon, has been prepared by the editor. In addition, those attending were invited to submit short contributions, and one or two of these are included in the written-up discussion.

Most of the points raised have thus been covered, either in the revised papers, or written-up discussion.

Some minor editing of the final versions of the papers has been done in order to achieve uniformity of presentation.

No attempt has been made to include chairmen's opening remarks, which, however, were usually very much to the point and invariably humorous!

The editor would like to thank Mr. James Hadley and the office staff of The Rank Prize Funds for their unfailing help, and also a number of his colleagues for assistance in deciphering the transcripts.

R. A. SMITH *Editor*

Contents

LIST OF PARTICIPANTS v

PREFACE ix

INTRODUCTORY REMARKS xv

Chapter 1 **High Resolution Atomic Spectronopy by means of Tunable Lasers**—P. JACQUINOT 1
    Stepwise excitation and 2-photon transitions . . . 6
    Comparison of 2-step and 2-photon transitions . . . 8
    Absorption line narrowing (A.L.N.) 8
    Coherent optical spectroscopy 10
    Perturbed atomic fluorescence experiments . . . 12

Chapter 2 **High Resolution Infrared Spectroscopy: The Spin-Flip Raman Laser**—S. D. SMITH 13
    Introduction 13
    Laser tuning mechanism 13
    Stimulated spin-flip Raman scattering 18
    Experimental 28
    Pulsed spectra 30
    Laser cavity properties of the SFRL 34
    cw SFRL spin-flip spectra 37
    Opto-acoustic spectroscopy with the SFRL: sensitivity of measurements 41
    Further developments 44
    Conclusions 47
    Acknowledgements 47
    References 48

Chapter 3 **Photon Correlation Spectroscopy**—E. R. PIKE . . . 51

CONTENTS

Chapter 4 **High Resolution Tunable Infrared Lasers**—A. MOORADIAN   75
    Semiconductor diode lasers 75
    Spin-flip Raman lasers 82
    Nonlinear mixing 87
    References 89

Chapter 5 **Generation and Measurement of Ultra-short Pulses**—D. J. BRADLEY 91
    References 110

Chapter 6 **Ultra-short Pulse Interaction Studies**—W. KAISER . . 111
    Acknowledgements 125
    General references 125

Chapter 7 **Coherent Optical Spectroscopy**—R. G. BREWER . . 127
    Introduction 127
    The Stark-switching technique 128
    Optical nutation 130
    Free induction decay and interference pulses . . 132
    Photon echoes and molecular collisions . . . 134
    References 142

Chapter 8 **Nonlinear Optical Techniques for Generation of VUV and Soft X-Ray Radiation**—S. E. HARRIS 143
    References 156

Chapter 9 **High Resolution Astronomy between Three Microns and Three Millimetres**—C. H. TOWNES 159
    References 185

Chapter 10 **Photoelectron Spectroscopy in the Study of Molecular Orbitals**—W. C. PRICE 187
    Introduction 187
    Photoelectron spectra of hydrides isoelectronic with the inert gases 189
    Photoelectron spectra of the halogen derivatives of methane . 194
    Spectra of ionic molecules 195
    Photoelectron spectra of "multiple" bonded diatomic molecules 198
    References 202
    General references 203

Chapter 11 **Spectroscopy Within the Cell**—BRITTON CHANCE, P. LESLIE
DUTTON, JOHN S. LEIGH 205
    Introduction 205
    Solvato and electrochromic responses of caroteroids and
        merocyanines 212
    Kinetic studies of caroternoid responses 212
    Summary 225
    References 225

Chapter 12 **Computer Techniques for Spectroscopy**—E. J. MILLETT . 227
    Introduction 227
    Computing loads in practice 228
        Computer calculation speeds 228
        Scanned line spectra 229
        Multichannel and transform spectroscopy . . . 233
        Band spectra 234
    Computer systems 234
        The dedicated minicomputer 235
        Multiple access computer systems 236
        The MRL hierarchical system 238
    References 242

Discussion 243

Line Narrowing by Time-biasing in Fluorescence—G. W. Series . 257

References 260

Where do we go from here? 261

Concluding Remarks 269

SUBJECT INDEX 271

INTRODUCTORY REMARKS

by Dr. F. E. JONES, M.B.E., F.R.S.

First of all, may I thank you all for the great warmth of response to our invitation to come to this symposium. All the speakers we invited to come, except one, accepted our invitation. We, who started organizing this conference a year ago, are really overwhelmed at the response from you all and are grateful to many of you who have travelled great distances to join us here.

Lord Rank, who died in 1972, had two great activities in this country: he continued to build up a flour-milling and bread business, which is now called Ranks Hovis McDougall. This supplies a lot of our nation's bread and flour. He also built up The Rank Organization, starting in films and now incorporating a wide range of activities, including radio and television, scientific instruments, precision engineering and hotels, as well as having a relationship with Xerox of the United States.

Lord Rank provided many charities during his lifetime, and the last of these, The Rank Prize Funds, came into being a short time before he died. The Rank Prize Fund for Opto-electronics is endowed with a million-and-a-quarter pounds, and the Committee which I chair was formed to see how best the funds could be used to further the subject of opto-electronics. We have had a number of ideas, and one of them, suggested originally by Professor Bradley, and which looked very good to us, was to sponsor a symposium in this field of Very High Resolution Spectroscopy. Our aims were to take note of the fact that, with the advent of the tunable laser and methods of producing very short pulses, techniques were available to give two or three orders more in resolution than previously. You will see that the theme of the symposium is that in the first morning there will be papers on what might be called high resolution in the spectral sense; in the afternoon there will be papers centred more on high resolution in the temporal sense— very short pulse techniques. On the second day we aim to call on the users to see what they are doing with very high resolution spectroscopy, and we have papers by distinguished speakers from the fields of astronomy, chemistry and biology, to tell us what results they are getting with the new techniques

INTRODUCTION

now available. In the afternoon there is one paper giving us a review of the techniques that are now available for spectroscopists in the data-handling fields. Thereafter there is a final session when other speakers who want to make a short contribution can do so; there will also be time for further discussion on the papers. We are also hoping to get going some discussion on the problems of the users that remain—things that they would like to do, and the possibility of the extension of the present techniques to take us on still further into very high resolution. So we have the spectral papers, the temporal papers, the users, and a session which might be called "where do we go from here?"

Chapter 1

High Resolution Atomic Spectroscopy by means of Tunable Lasers

by P. Jacquinot, Laboratoire Aimé Cotton, C.N.R.S. II, Bât. 505, 91405-Orsay, France.

Traditionally, spectroscopy was done by analysing the complex light emitted or absorbed by atoms or molecules, by means of optical instruments. Resolution was usually limited by these instruments and/or by lack of energy. Resolving power and energy throughput of optical instruments have been progressively increased: for instance Fabry-Perot Spectroscopy and Fourier Spectroscopy have been major steps in this development. So that now resolution is not mainly limited by instruments but by various causes of broadening such as Doppler broadening within the medium under study.

Besides spectroscopy by analysis of complex light another class of methods has grown. In these methods one observes the response of the atomic system to tunable monochromatic radiation scanned through the spectral range under study: this class of methods can be called "Tunable Monochromatic Radiation Spectroscopy" (TMRS). Of course it existed in principle before the advent of lasers, but it is only with lasers that it is possible to have in hand radiation with the required characteristics: sharpness, power, stability, tunability.

In fact, since the recent progress in tunable dye lasers it is now possible to have single mode cw operation. In these conditions the linewidth (of the order of 1 MHz) is far beyond the normal Doppler limit, and smaller than most of the natural atomic linewidths (this would not yet be true for many molecular transitions). The power is sufficient to saturate most of the transitions from ground atomic states. Stability can be made good enough to match the linewidth. Fine tunability is easy and coarse tunability is possible although not always very convenient. One point is still not very satisfactory. The wavelength coverage of cw dye lasers is still too small: Rhodamine 6G, which is by far the best dye, works only between 5700 and 6400 Å and although it is possible to extend the domain by using other dyes, this is not very easy and most of the experiments are still restricted to the range of Rhodamine 6G. When a pulsed laser can be used it is much easier to cover the domain from near ultraviolet to infrared. But it is reasonable to hope that much progress will be made in the near future.

We are thus able to have a resolution which is not limited by instruments or by the laser linewidth. But, of course, it can be limited by inhomogeneous broadenings, such as Doppler broadening, in the atomic medium itself. Fortunately there are several means of obtaining that the response of atoms to monochromatic excitation is determined only by the natural width of the levels, or, more generally by their homogeneous width. And this will be our definition of "High Resolution Spectroscopy": spectroscopy beyond the Doppler limit.

It may be instructive to represent the different methods of spectroscopy on the same diagram in order to understand the relationships between them. The most convenient representation I have found is a polar diagram in which the resolving power increases with the distance from the centre: the most external zone corresponds to our present definition of "High Resolution Spectroscopy". Angular *sectors*, or subsectors, correspond to different methods or techniques, and are grouped to show the parentage between them. For instance the left part of the diagram corresponds to "sequential" methods whereas the right part corresponds to "multiplex" methods; linear or nonlinear processes, or methods making use of coherent superposition of states are grouped together (Fig. 1).

We can now examine the different sectors of the diagram.

I will say very little about optical analysis of complex light. Let us only notice that with a very long spherical Fabry-Perot it is possible to study the linewidths of some lasers. Some high resolution work in atomic spectroscopy has also been made with atomic beams before the advent of lasers. Let us also recall that in Fourier Transform Spectroscopy, the signal delivered by a Michelson interferometer is the Fourier transform of the spectrum that would be recorded by a grating or Fabry-Perot spectrometer (assuming that overlapping of orders has been avoided). It is thus possible to say that Fabry-Perot or grating spectroscopy and Fourier Transform Spectroscopy are Fourier conjugates of each other: we shall see another example in the upper part of the diagram.

We now consider Tunable Monochromatic Radiation Spectroscopy methods. The simplest of all is obviously linear absorption of the light of a tunable laser by the medium under study. But since in gases or vapours the

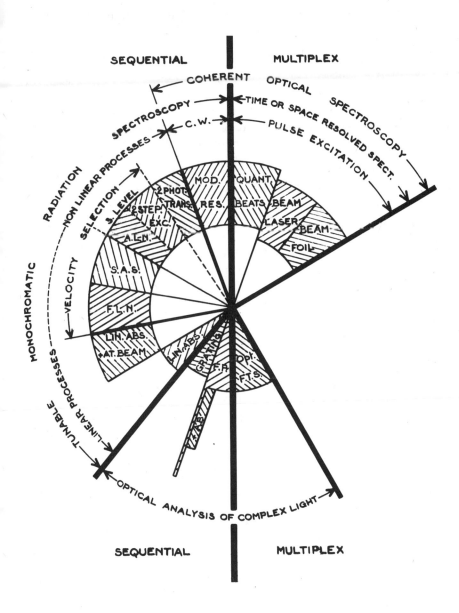

FIGURE 1. F.P. = Fabry-Perot Spectroscopy; F.T.S. = Fourier Transform Spectroscopy; A.B. = atomic beam; F.L.N. = (laser induced) Fluorescence Line Narrowing; A.L.N. = (laser induced) Absorption Line Narrowing; S.A.S. = Saturated Absorption Spectroscopy.

linewidth is limited by the Doppler effect, this is not High Resolution Spectroscopy in our sense and it is not within the scope of this talk. However it is worth saying that this will certainly become a very important method of spectroscopy with many advantages of resolution, signal-to-noise ratio, rapidity and simplicity.

If linear absorption is produced by an atomic beam, then Doppler broadening can be enormously reduced if the atomic beam is highly collimated and perpendicular to the light beam. Since in an atomic beam the atomic density is usually very low the absorption process cannot be monitored by the observation of the transmitted light: but observation of the fluorescent light from the upper level is a very sensitive means of detection of the absorption process, and it has been used in most of the experiments published so far. It is also possible to use other methods of detection like change in deviation of atomic trajectories in inhomogeneous magnetic fields (Rabi-type method), or deflection by recoil of the resonant atoms or photoionization from the upper level of the transition: these methods have been tested, they give the same resolution as detection by fluorescence and they should be more sensitive.

These experiments with atomic beams are not very difficult, and, in some cases it is even easier to use a beam than a vapour in a cell. Moreover an atomic beam is an ideal medium for the study of atomic properties because there are very few collisions and atoms can be considered as isolated. Provided the power of the exciting light is not too high there are no spurious effects as in many other methods and the interpretation of the results is simple and direct. However as the power increases such effects as power broadening or distortion of intensities of some components by optical pumping can occur.

Experiments have been published during the last two years. Owing to the limited wavelength coverage of dye lasers these experiments have been made mainly on sodium D lines. But it is also possible to study transitions starting from metastable states excited in the beam by electron bombardment, in noble gases for instance.

Hyperfine structures of the sodium D lines have been published by several laboratories. One of the best examples, shown in Fig. 2, has been obtained by a group in Hannover. The linewidth is of the order of 15 MHz, very close to the natural width of the excited level. The feasibility of atomic spectroscopy limited only by the natural width, in well controlled conditions, is thus demonstrated: its extension depends essentially on the extension of the wavelength coverage of dye lasers.

As an important by-product of this high resolution spectroscopy we have to mention different proposals recently made for isotope separation by means of the isotope shift. The principle of the proposed methods is very

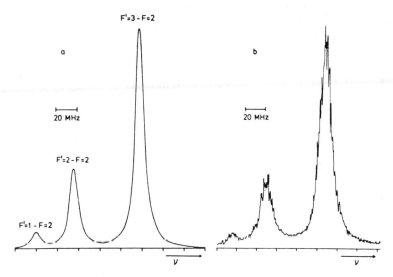

FIGURE 2. Part of the hyperfine structure of the D_2 line of sodium, obtained by recording the fluorescence of an atomic beam, excited by a tunable cw dye laser. (a) calculated signal, with natural line-width; (b) recorded signal.
(from Lange, W., Luther, J., Nottbeck, B. and Schröder, H. W. (1973), *Optics Communs* **8**, 157).

similar to the methods of nonoptical detection of optical resonances by means of beam deflection or photoionization from the upper level of the transition. For instance separation of barium isotopes has been recently reported by a group at Livermore, using beam deflection under the action of a laser beam at the resonance wavelength 5535 Å. In the case of calcium (at Cologne) and uranium (at Livermore) photoionization from a selectively excited state was used.

Atomic beams have also been used for the study of the light scattered by atoms excited by a radiation narrower than the natural width. Different theories exist which do not give the same results and it is of great interest to have experimental results. In an experiment by Schuda, Stroud and Hercher on sodium D_2 line, the scattered light was analysed by a high resolution Fabry-Perot interferometer: two Stark-effect sidebands, the splitting of which increases with the light intensity, were clearly seen.

In methods corresponding to the following sectors of the diagrams, named F.L.N. for (laser induced) Fluorescence Line Narrowing, S.A.S. for Saturated Absorption Spectroscopy, A.L.N. for Absorption Line Narrowing, and 2 Step exc. for 2-step transitions or excitation, the Doppler broadening is avoided by a velocity selection among the atoms of a gas: this selection is due to the fact that atoms are excited by the laser radiation only if they belong to a class of velocity such that, taking account of the Doppler shift, they are exactly tuned to the laser frequency, within their homogeneous width. This fact may be exploited in different ways corresponding to the different types of methods.

Fluorescence line narrowing is an interesting method which has been very little used. A gas is illuminated by a single mode laser beam and the fluorescence is observed in a direction opposite to the exciting beam. It is easy to see that, in these conditions, the structure of the levels excited within a Doppler linewidth appears in the fluorescent beam without Doppler broadening: this structure can be analysed by means of a high resolution Fabry-Perot spectrometer.

Saturated Absorption Spectroscopy is so well known that I believe I can omit it in this talk. Several excellent lectures have been given in several conferences and the importance of this technique and the spectacular results obtained have been stressed many times. So it will be sufficient for us to show the place of Saturated Absorption Spectroscopy in the general diagram.

We can now examine almost simultaneously two of the following sectors because they have many common features.

STEPWISE EXCITATION AND 2-PHOTON TRANSITIONS

These processes can be used for high resolution spectroscopy with a triple aim:
 (1) Possibility of reaching states not attainable by direct transition from the ground state because of selection rules;
 (2) Extension of high resolution spectroscopy to frequencies higher than those permitted by available tunable lasers;
 (3) Cancellation of Doppler broadening without the use of atomic beams.

For instance Fig. 3 shows the two possibilities for reaching 5S or 4D levels from the ground state 3S of sodium: "two-photon transitions" correspond to vertical arrows, whereas "two-step excitation" correspond to oblique arrows.

Two-step transitions obviously fulfill conditions (1) and (2). Condition (3) is also fulfilled if the two exciting beams are colinear: in this case, in effect, the first laser beam populates the intermediate level with atoms of only one class of velocity and only these atoms are available for the second transition: this is a velocity selection like in Fluorescence Line Narrowing or Saturated Absorption Spectroscopy. But spurious resonances occur if the intermediate level has several components within the Doppler width since these sublevels are populated by atoms belonging to different classes of velocity. It is easy to show that if there are two intermediate sublevels separated by δv there appears, in the case of two beams v_1 and v_2 propagating in opposite direction,

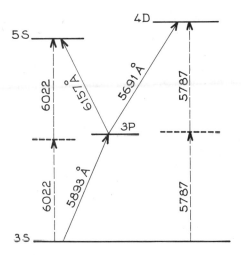

FIGURE 3. Two-step and two-photon transitions in sodium.

a spurious resonance at a frequency $v_2 = v_2 + \delta v(v_2/v_1 - 1)$: so these spurious signals would disappear at $v_2 = v_1$, and can effectively be unresolved from the main resonances for usual values of δv.

Up to now only one example of High Resolution Spectroscopy by means of a 2-step transition has been given on the 3S → 3P → 5S transition of Na. The experiment is simple in principle but it requires two single mode tunable dye lasers, one of which must be very precisely *stabilized* on the first transition. Like in single transition experiments the resonance was monitored by fluorescence. In this experiment the hyperfine structure of the 5S level has been directly measured.

In fact the experiment has been made with an atomic beam of Na, but only because this beam was available.

Two-photon transitions fulfill conditions (1) and (2) and also conditions (3) if the photons are of equal frequency and propagate in opposite directions. In that case there is an exact cancellation of the Doppler effect for all atoms regardless of the direction of their velocity so that all atoms participate in the phenomenon instead of only one class of velocity, as in the former case. Of course transition probabilities for these transitions are orders of magnitude weaker than in the former case. However very good signals have been obtained in several laboratories even with the moderate power of a few mW delivered by a cw dye laser; it is true that the transitions studied, 3S → 5S, 3S → 4D of Na were rather favourable because of the presence of the 3P level which is not too far from the middle of the two levels involved in the transitions (in the expression of the transition probabilities for this 2-photon process, there is a resonance denominator containing this distance). In one year four groups have published beautiful results on the hyperfine structure of the 5S and 6S levels, the fine structure of 4D and 5D levels, and also on the Zeeman and Paschen Back effect of 4D (Fig. 4). These results are one of the most spectacular achievements in the field of High Resolution Spectroscopy by means of tunable lasers and demonstrate the important potential of these methods.

COMPARISON OF 2-STEP AND 2-PHOTON TRANSITIONS

It is interesting to compare these two processes from a methodological point of view.

The advantages of the two-photon method are the following:
all atoms contribute to the signal: this is very important;
cancellation of Doppler effect is complete, without spurious signals;
only one laser is required, and it is easier to scan it through resonances than to stabilize one of the two lasers on one of the transitions.

But there are also some shortcomings:
in some cases the probability of the transitions may be very weak if there is no intermediate level well located;
owing to the necessity of using opposite beams with suitable polarizations in order to cancel the Doppler effect, the selection rules may severely restrict the number of observable components. For instance in the case of the above mentioned 3S → 5S transition the rule was $\Delta F = 0$ $\Delta m_F = 0$ so that only the difference between the hyperfine structures of 5S and 3S levels was measurable and Δv_{5S} was obtained by a difference between the measured quantity and the previously known Δv_{3S}.

In fact both processes are only two different particular cases of a same general phenomenon. For instance, it is possible to understand why, in the "2-step" process, only one velocity class of atoms contributes to the signal whereas in the "2-photon" process all atoms participate in the measurement, and to imagine a continuous transition from the first process to the second one. A rather elementary study of this point has been made but it would deserve a more complete analysis. An experiment has been recently reported (Liao and Bjorkholm, Bell Labs) on the resonant enhancement of two-photon transition, with two lasers of different frequencies when one of them approaches that of an intermediate state.

ABSORPTION LINE NARROWING (A.L.N.)

Like the two-step excitation this method works with a three-level system: when an intense monochromatic field interacts with one of two transitions sharing a common level, the absorption of the other transition is Doppler-

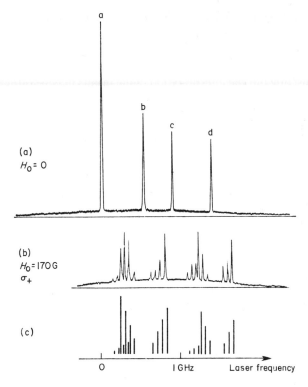

FIGURE 4. Zeeman effect of the two-photon transition 3S–4D of sodium. (c) calculated pattern (from Biraben, F., Cagnac, B. and Grynberg, G. (1974), *C. R. mebd. Séanc. Acad. Sci., Paris.* **B279**, 51).

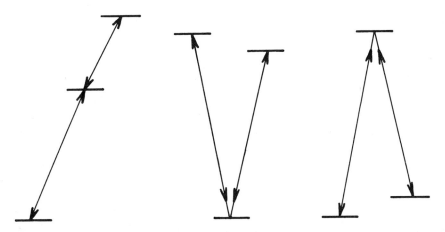

FIGURE 5. Different possibilities for coupled transitions in three-level systems.

free if it is observed in a direction colinear with the first beam. This also is an effect of velocity selection but it is more general than the 2-step transitions studied earlier since here the relative position of the three levels may be any one of those represented in Fig. 5. This effect may be used for High Resolution Spectroscopy but the interpretation of the results may be rather difficult.

COHERENT OPTICAL SPECTROSCOPY

This other class of methods makes use of the properties of a coherent superposition of quantum states.

It is well known that if an atom is prepared, by some suitable means, in a coherent superposition of excited states its radiation exhibits modulations, the frequencies of which are equal to the frequency differences between the coherently excited levels.

This preparation can be made by means of a short pulse of light tuned to the average position of the group of excited levels, the pulse being short enough so that its spectrum extends over the whole group. In this case one observes the free decay of the excited states: this decay shows the so-called "quantum beats". The Fourier transform of this decay gives the spectrum of the excited levels (see Chapter 7).

This preparation can also be made by means of a continuous optical wave modulated at a suitable frequency which matches the frequency difference between two levels of the group (more generally this could also be obtained by a coherent superposition of two waves with the suitable frequency difference). In this case the *continuous* fluorescence of the atom is modulated at the same frequency as the exciting beam and its *depth of modulation* has a resonant behaviour which permits the precise determination of the distances between the excited levels.

Of course these modulation frequencies are restricted to the range of radio frequencies so that only narrow structures (less than a few GHz) can be studied by this means. The same is true for quantum beats.

An important point is that the Doppler effect plays practically no role since the coherent waves which interfere are absorbed and emitted by the same atom so that the Doppler shifts affect only the low frequency of modulation and are then negligible. The linewidth is thus limited only by the homogeneous width of the levels even in a vapour.

In principle, lasers are not necessary for these experiments. And, in fact, quantum beats and modulation resonances have been studied and demonstrated in the early 1960's by Series and coworkers, and by Aleksandrov. But the availability of tunable lasers will make practical and useful these experiments which so far have been very difficult and limited to a very few favourable cases. In the case of pulsed excitation it is now possible to con-

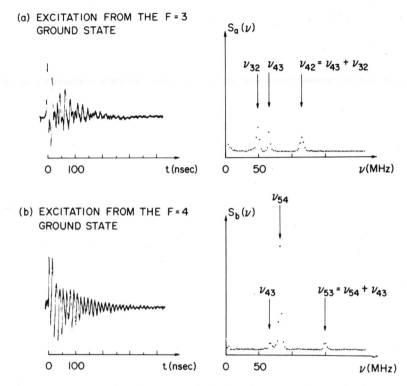

FIGURE 6. Fluorescence of caesium vapor submitted to short laser pulse tuned to the resonance line; left: the "quantum beats" recorded as a function of time; right: their Fourier transform. (from Haroche, S., Paisner, J. A. and Schawlow, A. L. (1973), *Phys. Rev. Lett.*, **30**, 948).

centrate enough energy in a very short pulse at the wanted wavelength. Whereas the quantum beats had, until recently, only been shown with Zeeman sublevels, beautiful experiments have been performed by Haroche, Paisner and Schawlow with hyperfine levels in caesium (Fig. 6).

In the case of modulation resonance the sharpness of the exciting line, given by a tunable laser and which is modulated, is by no means necessary. In that case it is only the intensity of the laser which will be important in order to obtain good signal-to-noise ratio. In fact this method of spectroscopy has almost never been used, even with lasers, and this is rather surprising.

It is obvious that modulation resonance spectroscopy and quantum beats spectroscopy are two different ways of making use of the same phenomenon and are Fourier conjugates of each other. A more complete analysis shows that the type of relationship between them is exactly the same as between grating (or Fabry-Perot) spectroscopy and Fourier Transform

Spectroscopy by means of a Michelson interferometer: the first is a sequential process and the spectrum is directly obtained by scanning; the other is a multiplex process and by taking the Fourier transform of the recorded signal one obtains exactly the same "spectrum" as by the first method. It is possible to develop the same considerations as for optical spectrometers concerning, for instance the multiplex advantage.

Very similar in its principle to quantum beats (or time resolved) spectroscopy is "space resolved" spectroscopy. In this method a fast ion beam is excited at a point in its path by a cw laser beam: this is equivalent to a pulse. Modulation in the decay is transformed into a space modulation along the direction of the ion beam. This "laser-beam" method has been applied with success by Andrä in Berlin.

The picture of optical coherent spectroscopy I have given here is far from being complete, for I have considered only the linear manifestations of interference between quantum states. But nonlinear effects may be important: they manifest themselves by a change in the *average* intensity of the fluorescent of absorbed light, in addition to modulation. A number of phenomena are related to this property.

PERTURBED ATOMIC FLUORESCENCE EXPERIMENTS

Another type of experiments should also be mentioned here, although the tunable laser is used only as a bright conventional tunable source. In these experiments recently developed on alkali atoms by Happer and Svanberg at Columbia University, an excited state is populated by a stepwise process—conventional resonance lamp plus tunable dye laser—and the well known usual effects of level crossing, Hanle effect or magnetic resonance are applied to this state: the occurrence of these effects is detected, as usual, by a change in intensity or polarization of the fluorescence from this excited state. This method combines the possibility of studying highly excited states and the advantages—and also some shortcomings—of radiofrequency or level crossing spectroscopy. These "perturbed atomic fluorescence experiments" (Happer) on highly excited states populated by tunable dye laser seem to have a very important potential for the knowledge of hyperfine structures in S and D states of alkalis along the Rydberg series.

Chapter 2

High Resolution Infrared Spectroscopy: The Spin-Flip Raman Laser

by S. D. Smith, Physics Department, Heriot-Watt University, Edinburgh, Scotland

INTRODUCTION

It is remarkable that it has taken us 300 years to make significant improvements upon the method of spectroscopy suggested by Newton in 1670. In Fig. 1, the conventional and recent methods are indicated schematically. The use of a broad-band black-body source, together with dispersing element is, of course, very inefficient in energy conversion and particularly so for the infrared region of the spectrum which occurs in the tail of the Planck energy distribution curve for high temperature black bodies. The infrared region around 10μm is about a million times worse off than the visible in this respect, and yields, to an order of magnitude, about 10^{-6} Watts per cm^{-1} frequency interval per square cm contained in a solid angle corresponding to an $f/5$ spectrometer. The effective transmission coefficient for spectrometers can be of the order of 10^{-2}–10^{-3} and effective powers for spectroscopic purposes as small as 10^{-10}W are therefore not uncommon.

LASER TUNING MECHANISM

From about 1970 onwards we have had new energy conversion mechanisms in tunable lasers as indicated on Fig. 2. It is of interest to extend the discussion of Jacquinot[1] concerning figures of merit for spectroscopic devices to include tunable lasers. Making a gross simplification to account for the diffraction-limited angular spread of the tunable laser beam, we may still make a useful comparison by considering the *spectral brightness*, i.e. the power per cm^{-1} of frequency interval, per steradian, per unit area. The

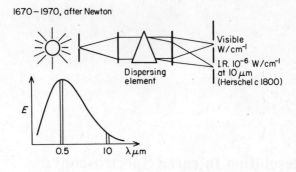

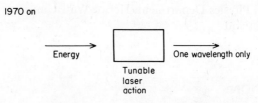

FIGURE 1. Schematic diagram of spectroscopic methods prior to, and since, 1970.

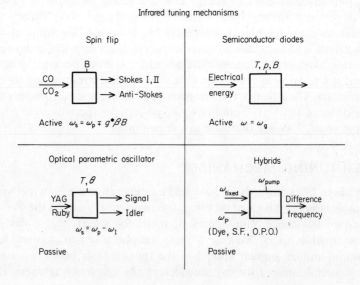

FIGURE 2. Principal mechanisms used for tunable infrared lasers since 1970.

estimated values for a number of tunable lasers will be discussed subsequently; the parameter enables one to calculate limits to frequency resolution, time resolution and sensitivity of spectroscopic measurements.

There have emerged a number of different mechanisms which give tunable laser outputs in the infrared part of the spectrum. The most useful of these are summarized in Fig. 2 in a diagrammatic form. At a time around 1964–65 the most obvious of these possibilities was the *optical parametric oscillator*. This makes use of the fact that a non-linear polarization can be created by a pump beam of frequency ω_p which has frequency components ω_i and ω_s such that $\omega_i + \omega_s = \omega_p$. These can be arbitrarily selected (that is tuned) by phase-matching techniques using temperature or angle as the adjusting parameter. Recent work has been reported from Stanford University on CdSe by Byer[2] and at Southampton University by Hanna et al.[3] on proustite for the range 1–20μm. So far, it is characterized by pulsed operation and rather moderate resolution.

A second method is the *semiconductor diode laser* (e.g. Melngailis[4] Blum and Nill[5], which differs from the first method in that the energy is introduced by means of an electric current instead of a pump laser beam. This creates population inversion between conduction and valence bands; electrons falling from the conduction band can generate stimulated radiation close to the frequency corresponding to the energy gap. The width of the gap can be tuned by temperature, pressure or magnetic field. This device has very high resolution and cw powers generally in the milliwatt range.

The third method is that of tunable stimulated Raman scattering. This paper will concentrate upon the process and application of *spin-flip Raman scattering*. It differs from the parametric oscillator in that the Raman scattering involves internal excitation of the crystal—the spin reversal of electrons in a magnetic field—and leads to laser output radiation which is magnetically tunable. It differs from the semiconductor diode laser in that instead of a population inversion, the Raman scattering process operates only when the lower state has a higher population than the upper state.

There exist other methods for the primary tuning of infrared lasers but we concentrate here on methods which have either given good spectroscopy, or show the promise of this. There exist also a number of hybrid systems which create infrared tunable radiation by mixing processes. Thus the visible dye laser as well as the spin-flip laser and optical parametric oscillator may be used to generate infrared radiation by difference frequency mixing.

With these various methods in existence, an attempt to make an estimation of comparative merit is inevitable. Such an attempt is made in Fig. 3 in which the criterion of spectral brightness is used. The first point to make is that in terms of this parameter nearly *all* tunable lasers are about a million times (or more) better than conventional spectroscopic methods. The numbers in

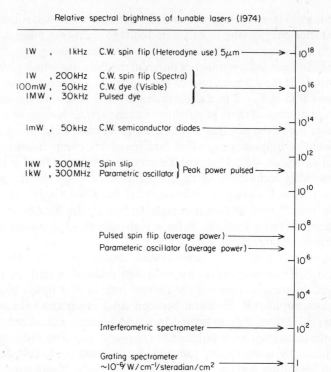

FIGURE 3. Relative spectral brightness of tunable lasers (1974).

the comparisons between different systems are certainly not better determined than to one or two orders of magnitude. They reveal, however, some interesting features, viz., that the cw systems, although of low power, at present show the biggest advantages because they have extremely narrow linewidths. At best, we are faced with the startling number of a possible factor of 10^{18} in the improvement of spectroscopic capability. One should also remark that the infrared has been very much the Cinderella of the spectral regions and it is only with these improvements that it is beginning to get on equal terms with visible and radio devices.

One aspect of the spectral brightness improvement that can be utilized is resolution and this is depicted in Fig. 4. Conventional grating spectrometers and interferometric spectrometers usually only reach linewidths of about 0.1cm^{-1}, which is not good enough to study even the pressure broadening of molecular transitions. Most of the tunable laser systems, even in the case of pulsed operation, can reach linewidths suitable to study pressure broadening and several systems, notably the semiconductor diode laser and the SFRL, have already reached two orders of magnitude below the Doppler

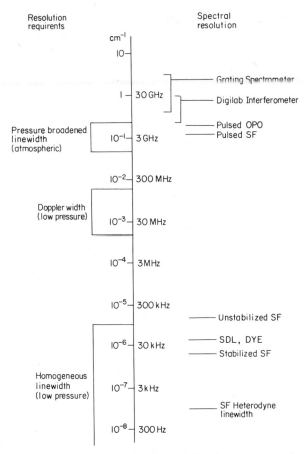

FIGURE 4. Resolution requirements and attainable resolution from conventional instruments and tunable lasers.

width (which can be seen as soon as pressures are reduced to about 1 Torr, and is of the order of 100MHz for most molecules). The ability to obtain linewidths well below the Doppler width then reveals such molecular effects as Λ-doubling, hot bands, isotope effects and in some favourable cases, nuclear hyperfine splitting. This is already making the infrared region competitive with other spectral regions.

Other ways in which we can make use of the spectral brightness advantages are, of course, in the sensitivity of the measurement or in the time resolution. It has also been demonstrated that sufficient power can be generated at linewidths less than the Doppler width to observe nonlinear effects in absorptions and hence by methods analogous to Lamb dip experiments to explore homogeneous linewidths.

STIMULATED SPIN-FLIP RAMAN SCATTERING

The Raman scattering process for the spin-flip excitation is not markedly different to the better known Raman scattering situations (Fig. 5). The incident photon frequency (ω_p) does not correspond to any transition between real states of the crystal; the inelastically scattered photon emerges in any direction (for spontaneous scattering) but the energy of the magnetic excitation given by the spin-reversal process is dependent upon the value of the applied magnetic field B, so that the scattered frequency ω_s, is given by:

$$\omega_s = \omega_p - g^* \beta B,$$

where β is the Bohr magneton, and g^* is the gyromagnetic ratio. In the material used, InSb, this has an anomalous value of around -50.

The Stokes frequency ω_s is thus tunable at a rate of $\sim 2\text{cm}^{-1}$ per kG or, more conveniently expressed for high resolution work, ~ 60MHz per gauss.

The Raman process has been understood quantum mechanically since at least 1928 (see, for example, Dirac[6]). Raman scattering is not just a direct transition process between two states; it is (like the polarization or refractive index) described by a two stage process in which a virtual transition is made to an intermediate state $|t\rangle$ followed immediately by a second virtual transition to a final state $|f\rangle$ which differs in energy from the initial state $|i\rangle$ by the excitation energy of the medium i.e. $\hbar g^* \beta B$.

The matrix element for the process is given by:

$$R = \sum_t \frac{\langle f|\varepsilon_s^* \cdot \mathbf{V}|t\rangle \langle t|\varepsilon_p \cdot \mathbf{V}|i\rangle}{\hbar\omega_p - (E_t - E_i)} - \frac{\langle f|\varepsilon_p \cdot \mathbf{V}|t\rangle \langle t|\varepsilon_s^* \cdot \mathbf{V}|i\rangle}{\hbar\omega_s + (E_t - E_i)} \quad (1)$$

where $\mathbf{V}$ is the electron velocity operator in the presence of a magnetic field. The resultant scattering transition probability is enhanced as the

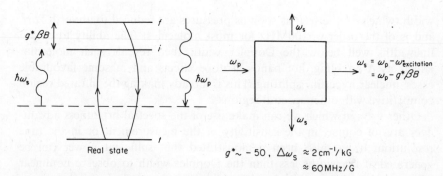

FIGURE 5. Schematic diagram of the second order spin-flip Raman process (left) and the experimental situation for spontaneous scattering (right).

incident photon energy ($\hbar\omega_p$) approaches ($E_t - E_i$) which is a resonant condition.

Truly *free-electron processes* give no magneto-Raman effects. Free-electron states in a magnetic field are given by Landau-like levels as shown in the upper part of the left-hand side of Fig. 6 and a magneto-Raman process could be envisaged as a spin-conserving transition between the $0 \to 1$ states followed by one between the $1 \to 2$ states. Since these are equally spaced the two terms of equation (1) will show exact cancellation since the energy denominators for the two terms are equal. There is no allowed transition with a change of spin quantum number either and so the

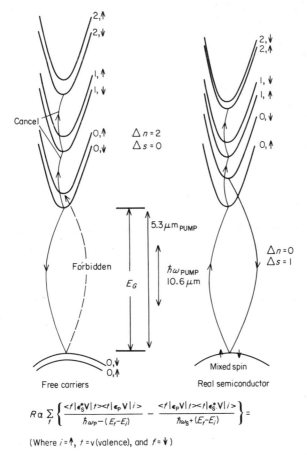

FIGURE 6. Schematic diagram of interband and intraband pairs of transitions giving rise to spin-flip Raman scattering. The idealized (parabolic band) situation on the left-hand side cannot give rise to this process. The mixed spin valence band, and non-parabolic conduction band, of a real semiconductor (right-hand side) are responsible for the scattering processes.

(interband) process is forbidden for free electrons. The fact that one can have several magneto-Raman effects in a real semiconductor is to be found in the detailed quantum mechanics of the states. Each step of the two transitions has to be electric dipole allowed and the left-hand diagram of Fig. 6 shows that this is *not* allowed for an *interband* transition for the simple energy bands of a semiconductor with an energy gap, E_G. For the cases of interest however, the effects of spin-orbit coupling lead to states at the top of the valence band having *mixed spin character* (see, for example Dennis et al.[7]). Transitions connecting the 0, ↑ state through the valence states to the 0, ↓ conduction state then become electric dipole allowed. This is the origin of the spin-flip scattering process. In addition, the small energy gap leads to strong interaction between conduction and valence bands and gives rise to nonparabolicity. For the magnetic states, this has the result that the $0 \to 1$ and $1 \to 2$ transitions no longer have equal energy and so an *intraband* process now becomes allowed as the two terms in equation (1) become unequal. Another result of this interaction is that the effective mass of conduction electrons becomes very small and, together with the spin-orbit splitting, causes the anomalous gyromagnetic ratio to become very large. The spin-flip and other magneto-Raman processes are therefore a special result of the detailed nature of the energy bands in these particular semiconductor crystals.

The history of the spin-flip Raman laser (SFRL) shows an interesting interaction between theory and experiment. This is summarized in Fig. 7. Wolff[8] theoretically predicted Raman scattering for the process $\Delta n = 2$, the intraband process referred to above, and in the same year Yafet[9] predicted the *spin-flip* process $\Delta s = 1$ via the *interband* transitions. Slusher et al.[10] observed both processes mentioned above and also an unpredicted process, $\Delta n = 1$, in spontaneous scattering.

To obtain laser action, one must satisfy the gain-loss relation

$$R \exp(g-\alpha)L > 0. \tag{2}$$

As shown in Fig. 7 the linewidth for the $\Delta s = 1$ spin-flip process was found to be much narrower than for the two orbital processes. This made the spin-flip process the most probable for laser action, since the gain g is proportional to the reciprocal of the linewidth, Γ, and is given by (e.g. Dennis et al.[7])

$$g = \frac{16\pi^2 c^2}{\hbar \omega_s^3 n_s^2} \cdot \frac{d\sigma}{d\Omega} \cdot \frac{I_p}{\Gamma} \tag{3}$$

where $d\sigma/d\Omega$ is the spontaneous Raman scattering cross section, and I_p is the pump intensity. Some considerable time elapsed before stimulated scattering was observed experimentally, however.

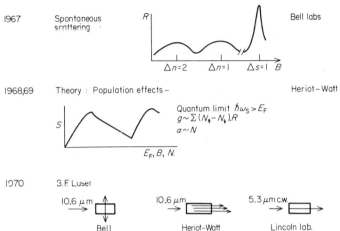

FIGURE 7. Review of the development of the spin-flip Raman laser from 1966 to 1970.

In 1968, from our work on interband Faraday rotation in InSb[11] we noted the analogy between the Raman scattering process, in terms of a nonlinear polarization, and the magneto-dispersion of Faraday rotation (a linear polarization). Both can be considered as two-step processes via an intermediate state. For Faraday rotation we had observed that, for interband effects near the energy gap, blocking of the intermediate state by filling the conduction band with electrons caused an oscillatory rotation rather similar to the de Haas-Van Alphen effect, depending upon the magnetic states rising out of the Fermi sea and becoming available for transitions. The analogy of this effect in spin-flip Raman scattering was reported by us at the Belfast Nonlinear Optics Conference in 1968 and the full calculation given by Wherrett and Harper[12]. The results are illustrated schematically in Fig. 7. To maximize the possibility of laser action, it is therefore necessary to adjust the final state to lie just above the Fermi level. Since the losses (expressed by the absorption coefficient α in equation 2) are linearly proportional to carrier concentration, N, favourable and unfavourable regions exist following the peaks of the oscillating Raman cross section. Thus, the first observation of spin-flip laser action by Patel and Shaw[13] was achieved at a magnetic field ~ 50kG so that $\hbar\omega_s > E_F$ with an electron concentration of 1×10^{16}cm^{-3}—called the "quantum limit" by Patel. Further work reported later in this paper show the effects of higher Landau levels and the effect is sometimes better described as "blocking" since there exists more

Spin-flip projects at Heriot-Watt, Edinburgh

5μm C.W.S.F.L. – Controlled spectroscopy of molecules. – Microwave – IR double resonance.	5μm + 10μm Pulsed S.F.L. – (time resolved) spectroscopy – Wide range, high powers
100μm S.F. Applications – Difference mixing in InSb. – S.F. pumping of molecular waveguide lasers.	Theory – Q.M. theory of S.F.R. scattering and difference frequency generation. – Rate equation theory of S.F.R.L.

FIGURE 8. Summary of spin-flip projects underway at Heriot-Watt University, Edinburgh.

than one quantum limit. This first observation was made with transverse scattering, thus facilitating the separation of the scattered radiation from the pump laser beam. However, it is more advantageous to use a colinear geometry since the spin-flip laser is then pumped over a much greater length and volume, i.e. that of the illuminated filament in the crystal. This geometry characterized the observations made in 1970 at Heriot-Watt University[14] and Lincoln Laboratory [15] shortly after Patel and Shaw's first report. The Heriot-Watt work showed also the simultaneous emission of second Stokes and anti-Stokes radiation, thus increasing the tuning range, and the Lincoln work utilized a 5.3μm cw CO laser which, with the aid of resonant enhancement resulting from near coincidence of $\hbar\omega_p$ with the energy gap of InSb, enables the SFRL to operate continuously.

This brief review of principles and progress up to 1970, together with Fig. 7, sets the stage from which we may describe the developments of the ensuing four years in greater detail. The research areas on which we have been concentrating in Edinburgh are summarized in Fig. 8. Apart from continuing to develop theory and experiment of the SFRL both for miscroscopic and macroscopic phenomena, we have put considerable effort into obtaining usable pulsed spectroscopy in the 5 and 10μm regions and cw spectroscopy over a wide range of resolution in the 5μm region. To obtain reliable spectroscopy has proved to be at least as difficult as discovering new effects for the first time but, as the spectra will show, progress is being made.

Figure 9 shows tuning curves obtained using a 10.6μm TEA CO_2 laser as a pump. The stimulated Stokes scattering gives a tuning range of about 160cm^{-1} for a magnetic field range from 30–100kG: the second Stokes process extends this a further 80cm^{-1}. The anti-Stokes process tunes in the opposite direction and also gives a range of about 160cm^{-1}. A total tuning range of about 400cm^{-1} is thus obtained using all three processes for one pump frequency at 945m^{-1}. More recently tuning has been reported[16]

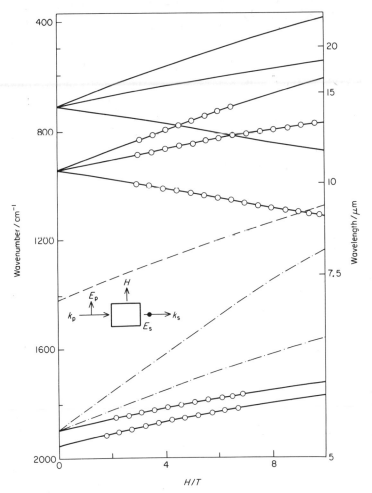

FIGURE 9. Tuning ranges available for the spin-flip Raman laser (circles). Additional ranges obtained either by use of a HF pump near 14 μm or with the aid of Non-linear Optics are also shown.

as high as 140kG, further extending this range. Peak power levels ~1kW were reported by Aggarwal et al.[17] for this type of excitation; in our own work peak powers are ~ a few hundreds of watts for Stokes radiation with conversion efficiencies of a few percent.

It is also possible to use the same pump laser via second harmonic generation to pump at 5.3μm[18, 19]. As indicated on Fig. 6, this frequency is then closely resonant with the intermediate states. It is readily possible to generate 5kW of second harmonic radiation using tellurium as a doubling crystal. This combination of relatively high power and resonant enhancement can be used to obtain spin-flip action under a wide variety of conditions—in particular low magnetic fields, pure samples and a wide range of pump wavelengths. Conversion efficiencies are now very high (70% or greater) and kilowatt powers in the 5.3μm range are readily obtained. Figure 10 shows how, by this method, with only one pump laser and a modest magnetic field of 7kG tuning can take place over more than 50cm^{-1}.

Frequency dependence of the SFRL output power for 10.6μm pumping is shown in Fig. 11 for Stokes, anti-Stokes and second Stokes processes[7]. This experiment provides one of the most direct ways of making direct comparison with microscopic theory. The onset of stimulated gain near

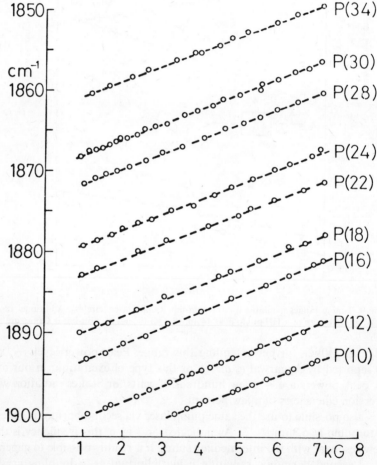

FIGURE 10. Tuning range available using a low field (7 k Gauss) spin-flip Raman laser pumped with frequency-doubled lines from a CO_2TEA laser near 5 μm.

30kG is associated with the first quantum limit of the blocking effect and we have identified the minima in the curve at higher fields as due to small absorption effects, i.e. α in equation (2) associated with phonon assisted harmonics of cyclotron resonance[20]. The exact positions of these peaks were identified in a separate absorption experiment carried out with the same pump laser operating at lower power. The final long-wave cut-off is caused by the tail of fundamental cyclotron resonance absorption being strong enough to overcome the gain. It is thus variable according to the

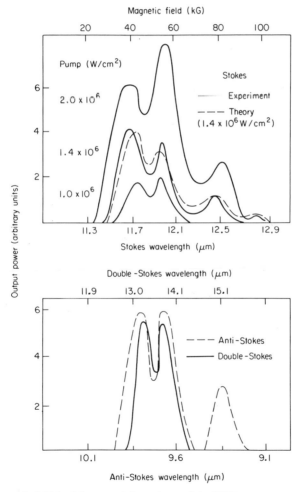

FIGURE 11. Magnetic field (and frequency) dependence of the SFRL output power for Stokes, anti-Stokes and second Stokes processing using 10.6 μm pumping. The dashed line above shows the theoretical curve calculated using microscopic semiconductor, and macroscopic laser, theory.

intensity of the pump, as shown in Fig. 11. The general form of this curve has been related to the theory in which we calculated the detailed matrix elements of all possible spin-flip transitions, shown in Fig. 12.[7,21] The theory also includes the necessary macroscopic rate equation [22,23] for a description of the growth of the stimulated wave in the spin-flip cavity in time and space. The complete theory (shown dashed on Fig. 11) is in good agreement with experiment and supports the basic microscopic assumptions of the spin-flip Raman process, together with the identification of the intermediate states. A spin-lattice relaxation time of 10nsec is inferred for a cavity with $N = 2 \times 10^{16} \text{cm}^{-3}$ and B of $\sim 50\text{kG}$.

The anti-Stokes and second Stokes processes are seen in the output curve of Fig. 11 to have similar form to the Stokes output. This almost certainly indicates that both processes are parametrically coupled to the Stokes wave.

In Fig. 13 we give some further experimental and theoretical results on the nature of the output against field for several different electron concentrations. The major maximum occurs when

$$\hbar\omega_s \approx E_F$$

in each case, showing how a smaller carrier concentration is more suitable for low field operation. For the case of 1×10^{16} electrons cm^{-3}, one may

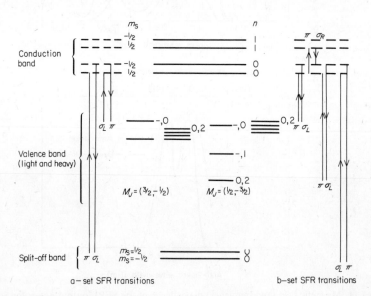

FIGURE 12. Schematic diagram of the conduction, valence and split-off band energy levels of InSb in a magnetic field (not to scale). The six pairs of transitions contributing to spin-flip Raman scattering are shown by vertical arrows.

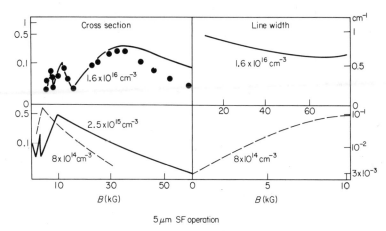

FIGURE 13. Experimental and theoretical results for the SFRL cross section and the spontaneous linewidth as a function of magnetic field and carrier concentration.

see on Fig. 13, peaks in the output and hence in the Raman scattering cross section occurring as higher Landau levels come through the Fermi level. These curves are a summary of both theory and experiment as presented by Wherrett et al.[24] at the Stuttgart Semiconductor Conference. The ability to tailor the carrier concentration for use at different fields is of practical importance and as shown in the right-hand side of Fig. 13 with curves constructed from data taken by Brueck and Mooradian[25] the spontaneous (and hence gain) *linewidth* depends both upon the concentration and upon the magnetic field. At high concentrations (greater than $10^{16} cm^{-3}$) the linewidth is nearly constant with field and of the order of $0.5 cm^{-1}$; this width may be greater than the axial mode spacing of a spin-flip laser cavity. At low concentrations, however, the linewidth varies with field and SFRL operation can readily be obtained with a gain linewidth as low as $10^{-2} cm^{-1}$—a usefully narrow linewidth for spectroscopy—and it changes almost by an order of magnitude for an increase of field of 10kG. Figure 13 summarizes the considerable degree of control that one has over the important laser parameters of gain and linewidth using the variables magnetic field and carrier concentration.

A typical SFRL input/output curve is shown in Fig. 14. The lowest threshold reported so far has been obtained with a 5mW output from the pump laser, while we normally operate with about 40mW for the cw case. This corresponds in intensity to $\sim 4 w/cm^2$. Input powers as high as 100kW have also been used and cross-sectional areas between a few 10's of microns in radius and a square centimetre. The pulsed burn threshold for InSb is around 20MW/cm² and there is therefore considerable scope for high power generation by using large-volume systems. A further interesting

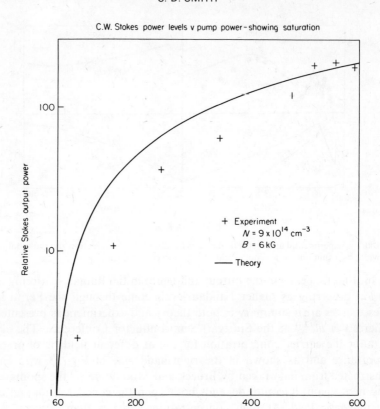

FIGURE 14. Typical cw input–output power curve (experiment and theory).

feature of the input/output curve is that with large power densities the output saturates. This is readily explained since it is possible to inject more pump photons than there are free electrons available. Thus all the spin states are excited and it is then necessary to await relaxation processes for the power to grow further. This feature is rather useful for spectroscopic purposes as it enables mode structure to be effectively removed in certain regimes of operation, albeit with some loss of resolution.

EXPERIMENTAL

The equipment for a spin-flip Raman laser (SFRL) is similar for both pulsed and cw operation. We illustrate diagramatically one of the forms in Fig. 15. The CO_2 TEA laser, giving a pulsed output of 100kW peak power and ~200nsec pulse duration, is weakly focussed on to a doubling crystal of tellurium of the order of 1cm³ in volume. A conversion efficiency of ~5%

of the 10.6μm radiation power allows approximately 5kW of 5.3μm radiation to be available for focusing on the spin-flip sample held in a cryostat between the pole tips of a conventional electro-magnet capable of generating a field of about 10kG. The temperature control of the spin-flip sample is rather important and the sample, of typical dimensions $8 \times 4 \times 4$mm, is usually soldered to the tail piece containing helium and thereby reaches a temperature of about 10°K. Up to 1kW of tunable Stokes radiation is readily obtained with quite high conversion efficiency. After rejecting the pump either with an interference filter or a simple grating monochromator, the tunable laser beam is passed through a gas absorption cell and a second channel is diverted by beam splitter to provide an intensity reference. We have used both fast Cu:Ge detectors and pyroelectric detectors for spec-

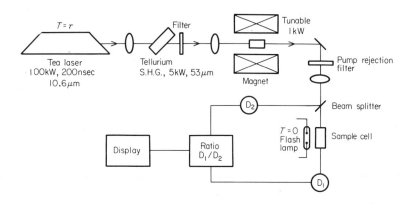

FIGURE 15. Schematic diagram of the SFRL double beam spectrometer.

troscopy with this SFRL. The signals are combined in an Edinburgh Instruments double channel synchronous ratiometer and can be displayed either as a ratio or simultaneously in each channel. This treatment is necessary in the case of pulsed operation due to pulse-to-pulse variation in the pump laser output. Double-beam presentation can also be useful in cw operation due to the pronounced mode structure in the output which will be discussed later in this paper. It is perhaps worth emphasizing that the entire spin-flip technology is within the scope of a typical university physics laboratory. Materials problems are not great, only liquid helium facilities complicate the technique, and even this problem may be solved in future by closed-cycle cryogenics. This is in contrast to, e.g. semiconductor diode lasers which require good semiconductor technology.

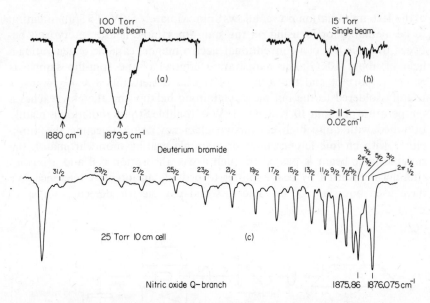

FIGURE 16. Molecular gas spectra taken with the pulsed SFRL near 5.3 μm. (a) DBr at 100 Torr pressure in a 20 cm long cell taken with the double beam system and fast detectors; (b) Single beam spectrum of DBr at 15 Torr pressure demonstrating the SFRL linewidth;[19] (c) Recent spectrum of NO Q-branch at 25 Torr pressure using a slow pyroelectric detector and Edinburgh Instruments' ratiometer.

PULSED SPECTRA

Figure 16 shows spectra taken with pulsed 5.3μm pumping. In the top left (a) we see the spectrum of deuterium bromide (DBr) at 100 Torr pressure in a 20cm long cell. The spectrum extends over about 1cm^{-1} and shows the isotope splitting between $Br^{78 \cdot 91}$ and $Br^{80 \cdot 91}$ in the $J = 4 \to J = 5$ R-branch line. The tunable laser linewidth is of the order of 0.02cm^{-1} and the signal-to-noise ratio is of the order of 100:1. The detectors were Cu:Ge photoconductive elements operating at 4K and the double-beam division was accomplished by gating within the 200nsec laser pulse. We term this "fast dividing". The remaining noise is mainly due to laser "jitter", not to detector noise. A conventional grating spectrometer would only just resolve these lines and the laser spectrum with peak power ~1kW is probably a thousand times better in terms of spectral brightness—well on the way to the predicted number of about 10^6. This demonstration of isotope line separation is interesting in itself and sufficient photometric accuracy was obtained to determine the isotopic abundances. The laser linewidth was determined by reducing the pressure and as shown in the single beam trace (Fig. 16 (b)), is limited to about the 0.02cm^{-1} or 600MHz. This work was reported at the Vail Conference on Laser Spectroscopy.[19]

The lower trace shows a more recent double-beam spectrum taken with similar pulsed excitation from a Gen-Tec TEA CO_2 pump laser operating at 40 pulses per second and again doubled in tellurium. The detectors on this occasion were two Mullard pyroelectric detectors with time constants of about 1msec. Thus unlike the technique in (a) the division, accomplished by an Edinburgh Instruments double-channel synchronous ratiometer, is integrated, over one pulse by the detectors and over several pulses in the dividing mechanism. We term this "slow dividing"; a considerable improvement in signal-to-noise ratio $\sim 25:1$ is obtained compared to single-beam operation at the same time constant. A spectral range of nearly 10cm^{-1} wide depicts the Q-branch of nitric oxide with a band head at 1876.075cm^{-1}. The signal-to-noise ratio is probably greater than 300:1. The indium antimonide spin-flip cavity had a carrier concentration of $8 \times 10^{14}\text{cm}^{-3}$ for this lower spectrum and the laser linewidth was again of the order of 0.02cm^{-1}. These curves give a good indication of the quality of pulsed spin-flip spectra that are presently possible. The total time during which the laser was radiating for these spectra was of the order of a few milliseconds and this indicates clearly that fast spectroscopy of considerable quality can now be achieved. Pulse-to-pulse instability of the pump laser is the main factor limiting performance.

The time bandwidth limit is around 0.003cm^{-1} and we hope that, with better pump lasers, even narrower tunable linewidths than 0.01cm^{-1} will be obtainable. With approximately kilowatt beam powers available in each pulse a rough calculation suggests that detection of 10^6 molecules in 1 microsecond pulse should be possible in principle. We are many orders of magnitude from achieving this in practice at present but the numbers may well become significant in the future.

Given pulsed outputs of the order of 1kW in peak power, it is readily possible to make nonlinear optical experiments using tunable infrared radiation. An early indication of this possibility was given in the work of Pidgeon et al.[26] who were able to add 10.6μm TEA CO_2 laser radiation to the tunable spin-flip output in the 10.8μm region to generate radiation of about 5.5μm wavelength. Another application is in difference mixing. In principle, a suitable nonlinear mixing crystal may be used to mix the pump and Raman-shifted beams resulting in far infrared radiation in the region of 100μm. A particularly interesting case is to mix these two beams together in the same material, indium antimonide. If we subtract the tunable Stokes frequency $(\omega_p - g^*\beta B)$ from ω_p, we obtain the difference frequency $\omega_3 = g^*\beta B$. This can therefore vary from $0-160\text{cm}^{-1}$ as B varies from 0–100kG. If the mixing is done in indium antimonide this *difference frequency* occurs exactly at the spin-flip frequency $g^*\beta B$. We have therefore a resonant three photon mixing process which can be described by the transitions indicated in

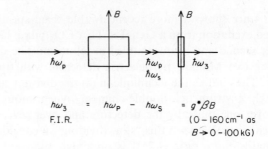
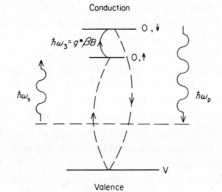

Far infrared difference mixing process

FIGURE 17. Schematic diagram of the experimental configuration and the transitions involved in the three-photon far-infrared difference mixing process.

Fig. 17. Two stages of the process are identical to the spin-flip Raman scattering and the power output at the difference frequency ω_3 is given by

$$P(\omega_3) \alpha \sum_t \sum_i R_t \frac{<\uparrow|\varepsilon_3 \cdot \mathbf{V}|\downarrow>}{[g^*\beta B - (\hbar\omega_p - \hbar\omega_s) + i\Gamma/2]},$$

where R_t is the matrix element previously calculated. The final step in the process is not purely electric dipole—the major contribution comes from a magnetization process. This resonant nonlinear process was first demonstrated by Van Tran and Bridges[27] using fixed 10.6 and 9.6μm lasers and then extended to mixing between the tunable spin-flip output and the pump.[28] We have repeated and extended this latter experiment using the

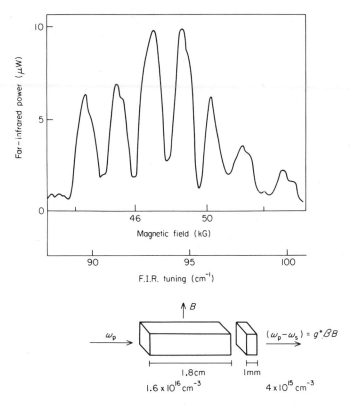

FIGURE 18. Far-infrared output power from difference mixing in InSb as a function of magnetic field (after Brignall et al.[29]).

greater power available from a TEA pulsed CO_2 laser as the pump source.[29] Figure 18 shows the results. Continuously tunable spectroscopy at 100μm is achieved; the pronounced peaks in the curve are, in fact, Fabry-Perot transmission fringes of the mixing crystal. Peak power levels of about 10μW were obtained, an increase of a factor of 5 on the earlier work at Bell Telephone Laboratories. The arrangement of spin-flip cavity and mixing crystal is shown in the lower part of Fig. 18. A carrier concentration of $1 \times 10^{16} cm^{-3}$ is used for the spin-flip crystal for maximum efficiency at 10.6μm while $4 \times 10^{15} cm^{-3}$ is used for the mixer in an attempt to maximize *phase-matching*. This has the effect of altering the *g*-factor for the two samples and there is a question as to whether the true resonance is obtained between the difference frequency and $g^* \beta B$, since both crystals are also in the same magnetic field of a superconducting magnet. It is possible that a small adjustment to the field for the mixing crystal could improve the output by at least a factor of 10. At this stage the tuning range is rather too limited and the power rather too low to make it a very practical device; considerable scope for development does, however, exist.

LASER CAVITY PROPERTIES OF THE SFRL

It will be noticed that the spectra of Fig. 16 exhibit apparently continuous tuning which although convenient for spectroscopy, suggests that full advantage of the true nature of the tunable laser output has not been seen. In Fig. 19 we diagrammatically indicate the circumstances for the SFRL. Cavities are typically ~ 10mm in length and this with a refractive index of 4.0 for InSb implies a separation for axial modes of $\sim 0.1 \text{cm}^{-1}$. The natural reflectance, 0.36, of the crystal is utilized and thus the transmission curve shows the normal characteristics of a low finesse Fabry-Perot etalon. This is shown in the centre of Fig. 19, together with the relationship between these cavity modes and the gain linewidths that can be adjusted with the aid of carrier concentration and magnetic field. Thus the gain linewidth can be made either much smaller or significantly larger than the axial mode separation. The SFRL can, therefore, operate in various different ways. For the case of narrow gain linewidth, it is possible to obtain operation on one axial

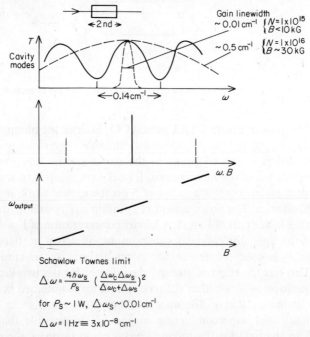

FIGURE 19. Schematic description of the effect of cavity modes on the tuning of the cw SFRL output with magnetic field.

mode and either from few or no transverse modes. The theoretical limit to the SFRL linewidth is then given by Schawlow-Townes expression:

$$\Delta\omega_{osc} = \frac{4\hbar\omega_s}{P_s}\left(\frac{\Delta\omega_c\Delta\omega_s}{\Delta\omega_c+\Delta\omega_s}\right)^2$$

Inserting appropriate quantities, this limit can be as low as 1Hz. In practice both the instability of the pump laser and the excitation of a number of transverse modes limit the linewidth to higher values. Around 10MHz is readily achieved with primitive equipment and 100kHz or less has been demonstrated by Patel,[30] passively, and ~30kHz by Brueck and Mooradian[25] using feedback techniques. The other extreme occurs when (say) 3 axial modes give laser action simultaneously; the linewidth is then of the order of 0.3cm^{-1}. In between these two extremes it is possible to operate with one axial mode and with results varying between 600MHz and ~1MHz —a very useful range for spectroscopy. This range of behaviour can be obtained by adjusting the input power and also the geometry of the beam. Figure 20 demonstrates the mode structure. In Fig. 20 (a) we see the gross

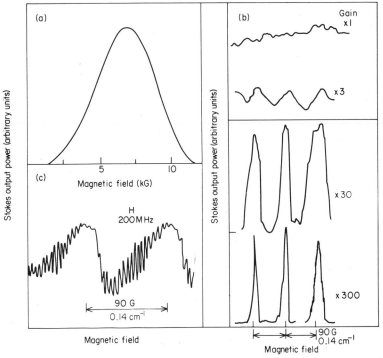

FIGURE 20. Mode structure of SFRL output in different regimes of operation. (a) Gross structure from 0–10 kG; (b) Form of output as the pump power is reduced by factors of 3, 30 and 300; (c) Fine detail of the individual modes.

structure as the magnetic field is varied from 0 to just over 10kG; the gain is effectively changing with field due to blocking effects; the pump laser power is high. At progressively decreasing powers, we obtain fine detail in the dependence of the output on magnetic field ranging from a relatively smooth curve at high powers to very pronounced mode structure at low powers when the SFRL switches off completely between modes. We can explain this behaviour in terms of the axial modes, which occur at a frequency separation of

$$\Delta v = 1/2nd \text{ cm}^{-1}$$

Now the SFRL gain occurs at frequencies given by $\omega_p - g^* \beta B$ and as the gain curve sweeps through the maxima of the cavity transmission, peaks of output occur; at low powers the laser may switch off. At intermediate powers, the cavity mode structure is apparent but the laser does not switch off and, finally, if the power is high enough we say the operation is in a "spin-saturated" regime in which the laser behaves as if the cavity modes were not present.

In even finer detail, Fig. 20(c), we may examine the structure of individual modes by using "frequency modulation" spectroscopy with an incremental oscillating magnetic field. This shows that there can be a wealth of fine structure on each axial mode; we have shown this to be due to external coupling effects.[40] It is, nevertheless possible by careful adjustment of gain and geometry to obtain very low-order operation of the SFRL and in these circumstances the resolution is very high.

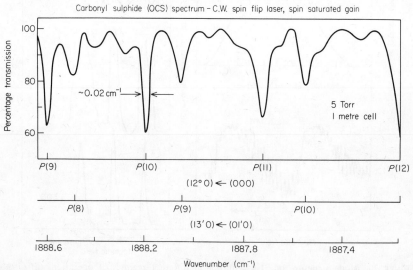

FIGURE 21. Spectrum of OCS near 1880 cm^{-1} taken with the cw SFRL in the spin-saturated regime.

cw SFRL SPIN-FLIP SPECTRA

We demonstrate spectroscopy under a variety of conditions for the cw SFRL. An early spectrum (1972) of the linear molecule OCS in a 1m path at a pressure of 5 Torr is shown in Fig. 21. We show about $1 cm^{-1}$ of spectrum near $1888 cm^{-1}$ of the $(12^00) - (00^00)$ transition of the rather weak combination band. The resolution is $\sim 0.02 cm^{-1}$ and well able to separate the basic structure controlled by the rotational constant $2B$. The cw SFRL was pumped with sufficient power to operate in the spin-saturated gain regime and the background was reasonably flat. A more recent transmission spectrum (1974) of the pyramidal molecule stibine, SbH_3, is shown

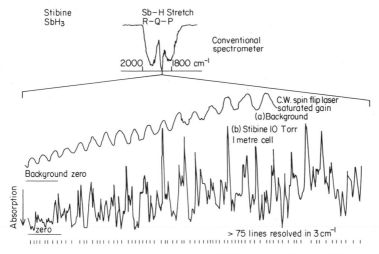

FIGURE 22. Transmission spectrum of part of the Stibine Q-branch near $1880 cm^{-1}$ (after Butcher et al.[31]).

in Fig. 22. In this case the gain is insufficient so saturate fully and the background shows about 10% modulation due to mode structure. We exhibit a small section from the middle of the Q-branch (shown in a spectrum from a conventional spectrometer, inset)—over 75 completely new lines were found—the laser linewidth on this occasion being $\sim 0.01 cm^{-1}$. There is a great deal of information available in this very complex spectrum; isotope splittings of antimony (^{123}Sb 42.75%) hot bands, and from comparison with millimeter-wave spectra of Jache et al.,[32] probably nuclear hyperfine structure as well. This is a good example of the power of the SFRL to obtain a very complex spectrum of a relatively simple molecule.

A favourite molecule to test the spectroscopic capability of tunable lasers is nitric oxide. It has an odd number of electrons and because of the Π electronic state exhibits P, Q and R-branches which lie conveniently in the range of the SFRL. Transmission spectra of the $Q(\frac{1}{2})_{\frac{1}{2}}$ line is shown in Fig. 23, obtained with the cw SFRL operating at low power (~100 milliwatts) and in the mode controlled regime. The large scale splitting (~600MHz) is due to Λ-doubling, i.e. the energy is slightly different according to whether Λ, the component of the electron orbital angular nuclear momentum, points in the O—N or N—O direction. Each line is split into two further lines because of ^{14}N nuclear hyperfine coupling, also known as quadrupole splitting. This splitting is of the order of 200MHz. The Doppler width of each of the four components is of the order of 125MHz and the pressure was only 175m Torr (1m cell). Doppler-limited SFRL spectroscopy is clearly demonstrated. The splittings can be estimated theoretically from ground state microwave data of Gallagher et al.[33] and Favero et al.[34] Using this information a crude tuning curve across the SFRL mode can be obtained. This is seen to be somewhat nonlinear[41],[42] and agrees well with a similar determination made by other methods by Brueck and Mooradian.[25] The

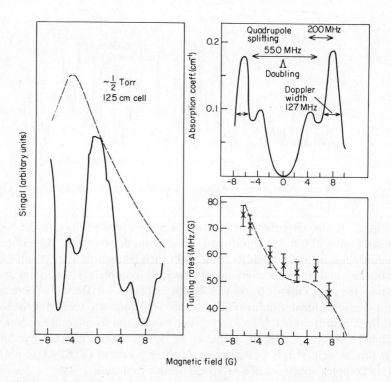

FIGURE 23. Absorption spectrum of the $Q(\frac{1}{2})_{\frac{1}{2}}$ line, of NO, illustrating Λ-doubling and nuclear hyperfine splitting. Also shown is a crude tuning curve obtained from the spectroscopic measurements.

maximum absorption coefficient ~0.036cm^{-1}/Torr is close to that obtained by Blum et al.[35] using a semiconductor diode laser. One should note that we see over 90% absorption which implies considerable spectral purity of the SFRL output and also that the effective laser linewidth is very much smaller than the Doppler width, perhaps in the range 1–10MHz.

The last three figures have summarized the spectroscopic behaviour of the cw SFRL as the pump power is steadily reduced, increasing the resolution down to below the Doppler limit. To discuss the new physics which this capability can uncover, it is of interest to note quantitatively the linewidths and shapes for pressure broadening, Doppler broadening and finally the natural linewidth of molecular transitions.

The usual Lorentzian expression for *pressure broadening* is

$$\alpha(v) = \frac{S}{\pi} \frac{\Delta v_p}{(v - v_0)^2 + (\Delta v_p/2)^2}$$

where α is the absorption coefficient, S the line intensity and the half width, Δv_p, about a centre frequency v_0, is

$$\Delta v_p = \Delta v_{p0} \cdot \frac{p}{p_0} \text{ but is unknown absolutely.}$$

We should note that conventional spectroscopy is unable seriously to examine the validity of this formula but from integrated absorption studies it is known that lineshapes are not well predicted (to a factor of the order of 3) at a distance of several half-widths from the centre. Experiments on pressure broadening give information on intermolecular forces, particularly at long range. We have made preliminary studies of pressure broadening in NO and OCS, two molecules which show a substantial difference in the rate of pressure broadening ($\approx$15MHz/Torr in NO and $\approx$5MHz/Torr in OCS). Detailed measurements have application in atmospheric transmission studies, e.g. the analysis that we require on the results from our Selective Chopper Radiometers on the Nimbus 4 and 5 satellites which require detailed line strengths, shape and widths for the 15μm Q-branch of CO_2 (see, for example, Abel et al.[36]). Calculations have already shown the inadequacy of Lorentzian lineshapes.

At pressures around 1 Torr the linewidth becomes *Doppler limited* and given by

$$\Delta v_D = (214/\lambda)(T/M)^{\frac{1}{2}} \text{MHz}$$

$$\alpha(v) = \frac{2S(\ln 2)^{\frac{1}{2}}}{\pi^{\frac{1}{2}} \Delta v_D} \exp\left(\frac{-4(v - v_0)^2 \ln 2}{(\Delta v_D)^2}\right)$$

This theory which gives $\Delta v_D = 127$MHz for NO and agrees well with experiments, both from this work with the SFRL and other work with semiconductor diode lasers. At this stage, reaching the Doppler width is mainly notable for revealing other finer molecular structure, such as the nuclear quadrupole splitting for NO mentioned above.

Under certain circumstances it is possible to eliminate Doppler broadening and so approach the *natural linewidth* given by *nonlinear* absorption:

$$\alpha_{NL}(\nu) \approx \alpha(\nu)\left[1 - \frac{2|\mu\mathscr{E}/\hbar|^2}{(\nu - \nu_0)^2 + (\Delta\nu_N/2)^2}\right]$$

$\Delta\nu_N = \dfrac{1}{\tau} \sim 10^{-1}\text{–}10^{-2}$ sec. and μ is the dipole moment.

Techniques include nonlinear saturated absorption, saturated fluorescence, two-photon absorption and quantum beat spectroscopy. The important feature from the point of view of the SFRL is now the relatively high power available. The electric field, $\mathscr{E}$, in the tunable infrared beam corresponds to a power of about 10 mW. As noted earlier, cw SFRL powers can readily be as high as 1W even for very narrow linewidths. A recent experiment[30] has demonstrated Lamb dip spectroscopy in water vapour; this is illustrated in Fig. 24. This shows resolution equivalent to an SFRL linewidth $\sim$100KHz for a passively stabilized system.

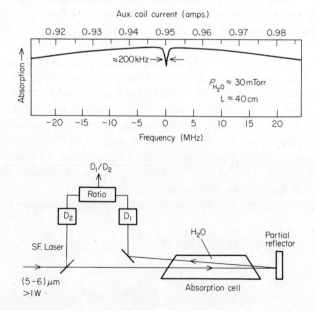

FIGURE 24. Lamb dip spectroscopy in water vapour taken with the cw SFRL (after Patel[30]).

OPTO-ACOUSTIC SPECTROSCOPY WITH THE SFRL: SENSITIVITY OF MEASUREMENTS

It is characteristic of most forms of tunable laser that at high resolution the cavity mode structure affects both the background spectrum and the frequency tuning characteristic. For the SFRL the background intensity problem is well illustrated in the lower curve of Fig. 25 where the spectrum of OCS is well nigh invisible against the cavity mode fluctuation. In addition to this, we need to be able to take advantage of the very much greater sensitivity. Although the signal-to-noise ratio is very much greater with the laser system such that detector noise is no longer a problem the laser intensity fluctuation sets the limit on photometric accuracy.

The answer to both these problems is a "modulation technique" in which only the absorption is recorded. A very useful technique is the opto-acoustic cell—originally invented before 1900 and applied first for use with tunable lasers by Kreuzer.[37] This device can be made extremely simply by inserting a side arm containing a commercial microphone in a normal absorption cell. This has a sensitivity of about 1 V/mTorr of gas pressure. The incoming tunable laser radiation is chopped and the signal from the microphone is subjected to synchronous detection to indicate the pressure change induced by absorption of the laser radiation. As the laser tunes through various absorption lines, the spectrum is exhibited. The upper curve of Fig. 25 shows

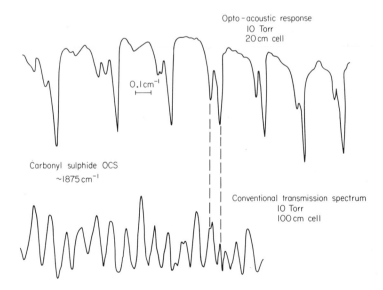

FIGURE 25. Comparison of opto-acoustic spectrum with conventional transmission spectrum of OCS.

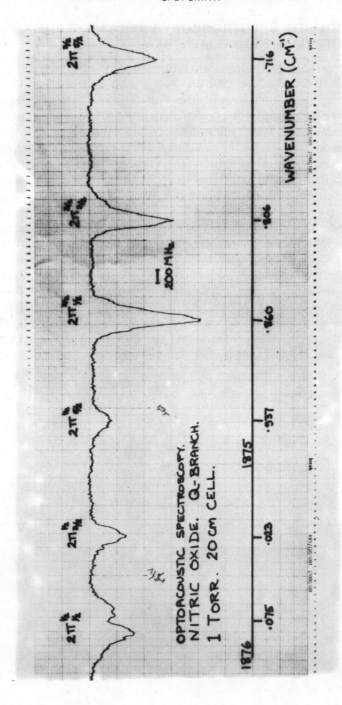

FIGURE 26. Photograph taken from a recorder trace of the opto-acoustic spectrum of part of a NO Q-branch.

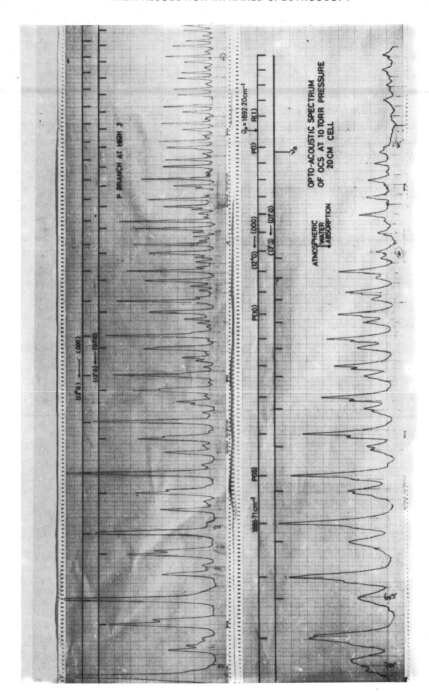

FIGURE 27. Wide range opto-acoustic spectrum of OCS

how the opto-acoustic cell extracts the small absorption signals to give a very clear spectrum for the OCS molecule. A further illustration of the quality of spectrum obtainable is shown in Fig. 26 which is an actual photograph taken from a recorder trace. With only 1 Torr pressure in a 20 cm cell a resolution of ~ 100 MHz was obtained with a signal-to-noise ratio of at least 100:1. Figures 25 and 26 show therefore that good quality SFRL, spectra can be obtained by this technique and for the purposes of finding line positions the combination of semi-saturated gain plus opto-acoustic detection is very straightforward and powerful. In terms of spectral brightness, these spectra are at least a million times better than conventional spectrometers can obtain. Figure 27 shows a wider range spectrum of OCS taken with the opto-acoustic cell. We have measured a complete P-branch and undertaken a band analysis using only SFRL data calibrated from a number of known frequencies. In general, calibration to about 0.01 cm^{-1} (300 MHz) is fairly readily achieved and Table 1 indicates the derived data for the $(13'0) \leftarrow (01'0)$ transition. The cw spectra presented here are more fully reported by Butcher et al.[31]

FURTHER DEVELOPMENTS

The main difficulties in progressing the SFRL as a spectrometer lie in the question of cavity mode control. We have already used temperature (by pumping over liquid helium) to shift cavity modes on to a molecular line but the technique is not very convenient. At this stage the cavity properties of the SFRL have been very little investigated. By developed laser standards a plane cavity of very short length is rather primitive. Good possibilities exist for external cavity operation, proper optical control of transverse modes and

TABLE 1. *Molecular Constants derived for OCS*

Transition	This Work[d]		Maki et al.[38]	Morino and Nakagawa[39]
	v_0 (cm^{-1})	(B–B)(10^5 cm^{-1})	v_0 (cm^{-1})	(B^1–B^{11}) (10^5 cm^{-1})
$(12^00)-(00^00)$[a]	1892.29 (0.01)	6.7 (1.4)	1892.20	9.49
$(13^10)-(01^10)$[b]	1891.685 (0.002)	16.2 (0.4)	1891.78	4.96
$(22^00)-(10^00)$[b]	1872.48 (0.01)	4.7 (0.8)	1872.45	16.9^0
$(14^00)-(02^00)$[b]	1889.640 (0.004)	2.7 (1.9)	1889.66	-3.9^0

a Using NO, H$_2$O calibration.
b Using OCS (12^{0}0)–(00^{0}0) calibration.
c From Maki et al.[39]
d Figures in parentheses are estimated standard deviations, which reflect the internal consistency of the data.

(13'0) ← (01'0) Transition

	P-Branch				R-Branch		
J	Field (kG)	Frequency (cm^{-1})	Obs − Calc (cm^{-1})		Field (kG)	Frequency (cm^{-1})	Obs − Calc (cm^{-1})
0	—	—	—		—	—	—
1	—	—	—		4.056	1892.501	+0.002
2	4.739	1890.901	+0.028		3.888	1892.895	−0.012
3	4.921	1890.473	+0.006		3.715	1893.302	−0.012
4	5.093	1890.070	+0.008		3.542	1893.708	0.014
5	5.266	1889.664	+0.008		3.366	1894.119	−0.011
6	5.440	1889.257	+0.005		3.184	1894.546	+0.007
7	5.614	1889.847	0.000		3.007	1894.961	+0.013
8	5.788	1888.440	−0.003		2.825	1895.388	+0.030
9	2.209	1888.039	0.000				
10	2.383	1887.633	−0.003		Standard error in a single frequency 0.025 cm^{-1}		
11	2.561	1887.219	−0.014				
12	2.733	1886.819	−0.011		(14^{0}0) ← (02^{0}0) Transition: P-Branch		
13	2.909	1886.407	−0.020	J			
14	3.083	1886.003	−0.022				
15	3.254	1885.605	−0.018	7	2.744	1886.793	0.000
16	3.427	1885.203	−0.019	8	2.920	1886.384	−0.002
17	3.599	1884.802	−0.018	9	3.099	1885.978	−0.002
18	3.767	1884.410	−0.009	10	3.270	1885.569	−0.004
19	3.935	1884.020	+0.001	11	3.445	1885.160	−0.007
20	4.103	1883.629	+0.011	12	3.618	1884.757	−0.003
21	4.274	1883.231	+0.013	13	3.791	1884.355	+0.001
22	4.446	1882.830	+0.011	14	3.964	1883.951	+0.003
23	4.618	1882.429	+0.009	15	4.135	1883.553	+0.011
24	4.788	1882.033	+0.012	16	4.310	1883.146	+0.011
25	4.959	1881.634	+0.012	17	4.487	1882.735	+0.006
26	5.132	1881.231	+0.007	18	4.661	1882.328	+0.005
27	5.308	1880.823	−0.003	19	4.838	1881.916	−0.001
				20	5.019	1880.495	−0.016
28	5.482	1880.416	−0.012		Standard error in a single frequency 0.007 cm^{-1}		
29	5.653	1880.018	−0.013				
30	—	—	—				
31	—	—	—				
32	4.251	1878.850	+0.010				
33	4.426	1878.445	+0.001				
34	4.596	1878.051	+0.002				
35	4.765	1877.659	+0.006				
36	4.936	1877.236	+0.005				
37	5.109	1876.862	−0.001				
38	5.282	1876.459	−0.010				
39	5.449	1876.072	−0.003				

piezoelectric control of cavity length. It would then be possible to feed back an error signal to the magnetic field control and so to place the $g^*\beta B$ tuning exactly on top of the cavity mode. By this method a large linear tuning range at very high resolution may be achievable. Further possibilities exist of using the magneto-optical properties of InSb and the directional nature of the spin-flip Raman scattering to provide an independent magneto-optical control of cavity length. Preliminary experiments using pulsed pump beams suggest that the effect is large enough to be seriously considered. Figure 28 summarizes some of these possibilities.[40]

The SFRL cavity operates using only the natural reflectivity (0.36) of InSb. Under the circumstances where the gain is high, this is not inappropriate but, as the discussion on blocking and quantum oscillations given earlier shows,

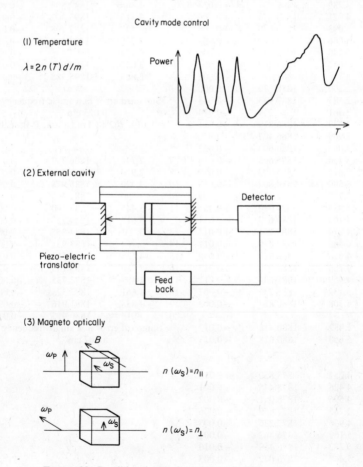

FIGURE 28. Possible schemes for achieving cavity mode control.

a very considerable degree of control may be affected on the gain. Since gain can be reduced by going to lower carrier concentrations, the absorption in the cavity is also reduced. Under these circumstances high efficiency dielectric coatings may be useful. This could lead to considerable increases in the range in which the InSb can be pumped and possibly to cw operation in the 10 μm region. Recently a second very promising material, CdHgTe, has given spin-flip laser action when pumped at 9.3 μm.[43] This is an alloy material with an adjustable energy gap and is likely to extend further the range over which SFRL action can be used for spectroscopy.

CONCLUSIONS

The microscopic understanding of the SFRL seems, at this stage, to be satisfactory. The two main effects, resonant enhancement and the blocking effect of electron population are quantitatively well understood. This enables parameters such as gain and linewidth to be adjusted in a controlled manner.

By contrast, the macroscopic physics of the SFRL although quite well developed theoretically, has been very poorly tested experimentally. This is mainly due to the difficulty of defining invisible infrared beams and the relatively poor quality of pump lasers. As a result of this situation only very primitive spin-flip laser cavities have so far been used.

Notwithstanding the last comment, we have demonstrated good quality spectroscopy in the resolution range 300 MHz to 10 MHz and further experiments have shown the SFRL's capability down to at least 100 kHz. The available powers in both cw and pulsed operation are sufficient to conduct nonlinear experiments necessary to go below the Doppler limit.

Figures of merit defining improvement in spectroscopy over conventional means have been demonstrated for the SFRL to be greater than 10^8, using equipment which can be developed and used in typical university physics and chemistry laboratories. Further developments in both power, resolution, stability and tuning range are likely and give great promise to this technique.

ACKNOWLEDGEMENTS

Much of the work reported here was carried out by the Spin-Flip Group at Heriot-Watt University over the period 1970–1974.

I wish in particular to acknowledge the contribution of Dr. M. J. Colles, Dr. R. B. Dennis, Dr. C. R. Pidgeon, Dr. R. A. Wood, Dr. A. McNeish, and T. Scragg on the experimental side, and of Dr. W. J. Firth and Dr. B. S. Wherrett on the theoretical side. Dr. R. L. Allwood, now of Edinburgh Instruments Limited, contributed as a member of the group in the earlier stages and latterly in cooperation with his Company. Dr. R. J. Butcher, of the Department of Physical Chemistry, University of Cambridge helped greatly with the molecular spectroscopy during the year 1973/74 which he spent in Edinburgh. Financial assistance from the Science Research Council and Heriot-Watt University is also gratefully acknowledged.

REFERENCES

1. Jacquinot, P. (1954), *J. Opt. Soc. Am.* **44**, 761.
2. Byer, R. L. (1973), *in* "Laser Spectroscopy", Brewer, R. G. and Mooradian, A. eds., pp. 77–101, Plenurn Press, New York.
3. Hanna, D. C., Luther Davies, B., Smith, R. C. (1973), *Appl. Phys. Lett.* **22**, 440.
4. Melngailis, I. (1973), *in* "Laser Spectroscopy", Brewer, R. G. and Mooradian, A. eds., p. 237, Plenum Press, New York.
5. Blum, F. A. and Nill, K. W. (1973), *in* "Laser Spectroscopy", Brewer, R. G. and Mooradian, A. eds., p. 493, Plenum Press, New York.
6. Dirac, P. A. M. (1970), The Principles of Quantum Mechanics, 4th Edition, Oxford University Press.
7. Dennis, R. B., Pidgeon, C. R., Wherrett, B. S., Wood, R. A. (1972), *Proc. R. Soc.*, **A331**, 203–236.
8. Wolff, P. A. (1966), *Phys. Rev. Lett.*, **16**, 225.
9. Yafet, Y. (1966), *Phys. Rev. Lett.*, **152**, 858.
10. Slusher, R. E., Patel, C. K. N., Fleury, P. R. (1967), *Phys. Rev. Lett.*, **18**, 77.
11. Dennis, R. B., Smith, S. D. and Summers, C. J. (1971), *Proc. R. Soc.*, **A321**, 303–320.
12. Wherrett, B. S. and Harper, P. G. (1969), *Phys. Rev.*, **183**, 692.
13. Patel, C. K. N., Shaw, E. D. (1970), *Phys. Rev. Lett.*, **24**, 251.
14. Allwood, R. L., Devine, S. D., Mellish, R. G., Smith, S. D. and Wood, R. A. (1970), *J. Phys. Chem. Solid St.*, **3**, L186.
15. Mooradian, A., Brueck, S. R. J., and Blum, F. A. F. A. (1970), *Appl. Phys. Lett.*, **17**, 481.
16. Grisar, R., and Wackernig, H. (1974), *Proc 12th Int. Conf. Phys. Semiconductors*, Teubner, Stuttgart, pp. 803–807.
17. Aggarwal, R. L., Lax, B., Chase, C. E., Pidgeon, C. R., Limbert, D., and Brown, F. (1971), *Appl. Phys. Lett.*, **18**, 383.
18. Allwood, R. L., Dennis, R. B., Smith, S. D., Wherrett, B. S., and Wood, R. A. (1971), *J. Phys. Chem.*, **4**, L63.
19. Smith, S. D., Pidgeon, C. R., Wood, R. A., McNeish, A., and Brignall, N. (1973), "Laser Spectroscopy", Brewer, R. G. and Mooradian, A. eds., Plenum Press, New York.
20. Dennis, R. B., Wood, R. A., Pidgeon, C. R., Smith, S. D., Smith, J. W. (1972), *J. Phys. Chem., Solid State Physics*, **5**, L73–79.
21. Dennis, R. B., Firth, W. J., McNeish, A., Pidgeon, C. R., Smith, S. D., Wherrett, B. S., and Wood, R. A. (1972), *Proc. 11th Int. Conf. Phys. of Semiconductors*, Warsaw.
22. Firth, W. J. (1972), *IEEE J. Quantum Electron.*, **8**, 865.
23. Wherrett, B. S. and Firth, W. J. (1972), *IEEE J. Quantum Electron.*, **8**, 869.
24. Wherrett, B. S., Wolland, S., Pidgeon, C. R., Dennis, R. B., and Smith, S. D. (1974), *12th Int. Conf. Phys. Semiconductors*, Stuttgart.
25. Brueck, S. R. J., and Mooradian, A. (1974), *IEEE J. Quantum Electron.*, **10**, 634.
26. Pidgeon, C. R., Lax, B., Aggarwal, R. L., Chase, C. E., and Brown, F. (1971), *Appl. Phys. Lett.*, **19**, 333.
27. Van Tran, Nguyen, and Bridges, T. J. (1972), *Phys. Rev. Lett.*, **29**, 359.
28. Van Tran, Nguyen, and Bridges, T. J. (1973), "Laser Spectroscopy", Brewer, R. G. and Mooradian, A. eds., Plenum Press, New York.
29. Brignall, N., Wood, R. A., Pidgeon, C. R., and Wherrett, B. S. (1974), *Optics Communs*, **12**, 17.
30. Patel, C. K. N. (1974), *Appl. Phys. Lett.*, **25**, 112.

31. Butcher, R. J., Dennis, R. B., and Smith, S. D., The Tunable Spin Flip Raman Laser, II: Continuous Wave Molecular Spectroscopy. Communicated, proceedings of Royal Society, 1975.
32. Jache, A. W., Blevins, G. S., and Gordy, W. (1955), *Phys. Rev.*, **97**, 680.
33. Gallagher, J. J., Beddard, R. D., and Johnson, C. M. (1954), *Phys. Rev.* **93**, 729.
34. Favero, P. G., Mirri, A. M., and Gordy, W. (1959), *Phys. Rev.*, **114**, 1534.
35. Blum, F. A., Nill, K. W., Calawa, A. R., and Harman, T. C. (1972), *Chem. Phys. Lett.*, **15**, 144.
36. Abel, P. G., Ellis, P. J., Houghton, J. T., Peckham, G. E., Rodgers, C. D., Smith, S. D., and Williamson, E. J. (1970), *Proc. R. Soc.*, **A. 320**, 355–55.
37. Kreuzer, L. B. (1971), *J. Appl. Phys.*, **42**, 2934.
38. Maki, A. G., Plyler, E. K. and Tidwell, E. D. (1962), *J. Res. Nat. Bureau of Standards*, **66A**, 163–167.
39. Morino, Y., and Nakagawa, T. (1968), *J. Mol. Spectrosc.*, **26**, 496–523.
40. Scragg, T., and Smith, S. D. (1975), *Optics Communs*, **15**, 166–168.
41. McKenzie, H. A., Smith, S. D., and Dennis, R. B. (1975), *Optics Communs*, **15**, 151–156.
42. Firth, W. J., Wherrett, B., and Weaire, D. L. (1975), *Optics Communs*, **15**, 157–160.
43. Weber, R. A., Sattler, J. P., and Nemarich, J. (1975), *Appl. Phys. Lett.*, **27**, 93.

Chapter 3

Photon Correlation Spectroscopy*

by E. R. Pike, Royal Radar Establishment, Malvern, Worcs. WR14 3PS, England

In this talk I shall outline the principles and applications of a new field of high resolution spectroscopy, which has become possible with the advent of the laser, based on analysis of the statistical properties of the detections of individual radiation quanta from the electromagnetic field. The work originated in fundamental statistical studies of laser light and other light sources and has now progressed to the point where it has developed into a very powerful practical tool in use in many fields of science for the study of molecular and macroscopic motion.

Let us start very generally by assuming that we have a source of some unspecified matter radiating out into space, and we want to describe the effects due to that source. We use a complex wave field to describe the radiation, and we place a detector in the field with an area A and response time T. The conventional quantum theory of photodetection tells us that we can define a quantity which we can loosely call the intensity; this is not a classical intensity in the exact sense, but a probability that one will observe a photoabsorption at the detector over its area and during its response time. This quantity $I(t)$ is simply constructed from the wave field in the following way:

$$I(t) = \int_A \int_t^{t+T} <\hat{\mathscr{E}}^+(\mathbf{r}, t') \hat{\mathscr{E}}^-(\mathbf{r}, t')> dS dt' \qquad (1)$$

The $\hat{\mathscr{E}}(\mathbf{r}t)$ are second-quantized field operators and the brackets indicate the quantum statistical average. Unfortunately the photon has a zero mass and spin one; this implies that there are unusual localization considerations—

*The material in this chapter is published by consent of the Ministry of Defence.

they are not really problems because the theory is understood—but one cannot localize a photon in space and time. However, as long as the detector area and the response time obey the relations:

$$A \gg \lambda^2, \quad T \gg \frac{1}{\omega_{\text{opt}}} \tag{2}$$

where λ is the optical wavelength and ω_{opt} the optical frequency, then we can essentially localize the photons; these conditions are very easily met in most of our experiments. The form of equation (1) is reminiscent of the probability of finding an electron at $(\mathbf{r}, t)$ given by Born's modulus squared of the electron wave function; the limitations (2), however, do not apply to a massive particle.

Let us now ask "What do we mean by the spectrum observed by the detector?". We shall answer this by means of an imaginary experiment. Let us take an infinite optical grating and define the spectrum by what is seen by the detector at a given angle and in a given order in the far field. What comes off the grating is a set of wave fields which are delayed from one another by a fixed delay τ (see Fig. 1).

In the far field we have the sum over the various grating components of a set of delayed fields

$$\hat{\mathcal{E}}(t, \tau) = \sum_{j=-\infty}^{+\infty} \hat{\mathcal{E}}(t + j\tau)$$

The detector response is thus

$$I(t, \tau) = \sum_{j,k} \int_A \int_t^{t+T} <\hat{\mathcal{E}}^+(t' + j\tau)\hat{\mathcal{E}}^-(t' + k\tau)> \, dS \, dt'$$

For a stationary field with sufficiently small A and T (while not violating the conditions of equation (2)) this becomes

$$I(\tau) = \text{Re} \sum_{\ell=-\infty}^{+\infty} <\hat{\mathcal{E}}^+(0)\hat{\mathcal{E}}^-(\ell\tau)>$$

which, using the result $\sum_{-\infty}^{+\infty} e^{inx} = 2\pi\delta(x - 2\pi\ell)$, gives

$$I(\tau) = \text{Re} \sum_{n=-\infty}^{\infty} \int_{-\infty}^{+\infty} <\hat{\mathcal{E}}^+(0)\hat{\mathcal{E}}^-(\tau')> e^{2\pi i n \tau'/\tau} \, d\tau'$$

The first-order spectrum ($n = 1$) is thus, writing ω for $2\pi/\tau$,

$$S(\omega) = \text{Re} \frac{1}{2\pi} \int_{-\infty}^{+\infty} G^{(1)}(\tau') e^{i\omega\tau'} \, d\tau'$$

$$= \text{Re} \quad (\text{Fourier transform of } G^{(1)}(\tau))$$

where $\quad G^{(1)}(\tau) = <\hat{\mathcal{E}}^+(0)\hat{\mathcal{E}}^-(\tau)>$

is the first-order field correlation function.

Let us suppose that the wave field has two frequencies present, f and $f + \Delta f$. (We are beginning to talk in spectroscopic language here because we shall take a frequency difference of 1 part to the 10^n—that is $\Delta f/f = 10^{-n}$.) Suppose we wish to measure the difference between these two frequencies with a classical optical instrument, such as a grating or Fabry-Perot interferometer, then clearly we have to create two paths within the instrument which will allow us to superimpose physically two waves which have at least one wavelength difference in their paths. This will tell us how long the apparatus has to be in order to do this experiment.

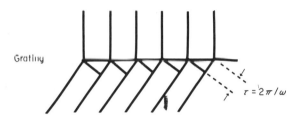

FIGURE 1. Experimental definition of the spectrum of a light field.

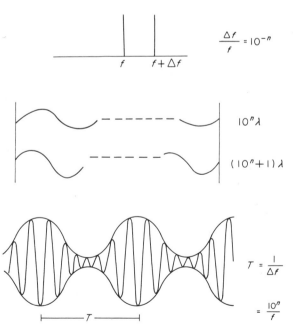

FIGURE 2. Scales of length, time and frequency for two-mode spectrum.

TABLE 1. Scales of length and time for instruments of various optical resolving powers.

Resolution	L	T
$1:10^5$	5 cm	200 psec
$1:10^6$	50 cm	2 nsec
$1:10^7$	5 m	20 nsec
$1:10^8$	50 m	0.2 μsec

Alternatively, we can actually just look at the response of the detector or the superposition of the two fields, and we will see that it has a beat component with cycle time T given by $T = 10^n/f$, (see Fig. 2). In Table 1 these relations are quantified. We see that as we increase the length of an instrument required to get higher and higher resolution we are actually reducing the time between the fluctuations down to times which are easily measured on a laboratory scale. By the time one is asking for resolutions around 1 in 10^8 then clearly one should wonder about using the time scale of the fluctuations rather than the length scale.

We see then that a broad spectrum will fluctuate quickly and a narrow spectrum will fluctuate slowly, the fluctuations being due to the various spectral components beating together.

Going back to equation (1) what one really measures when one places a detector in a fluctuating light field is the probability of the detector absorbing a photon at a particular time. What actually comes out of one's detector is a train of photodetections in time. This is illustrated in Fig. 3. An actual

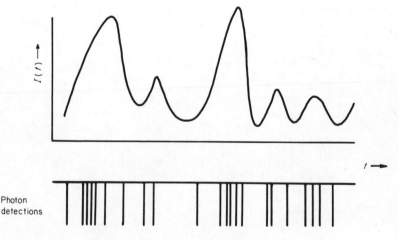

FIGURE 3. The response of an optical detector.

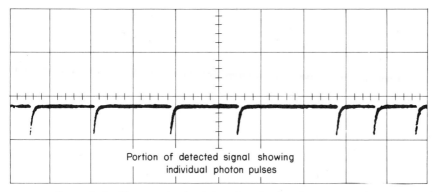

FIGURE 4. Oscillogram of train of photodetection impulses, 1 cm = 20 ns.

oscillogram of such a train of photodetection impulses is given in Fig. 4. This shows the response of a modern optical detection system consisting of a high quality phototube with a fast amplifier and pulse standardizing circuit. So really this is the form of the basic data of any optical experiment. It has only been apparent in this essentially ideal form in recent years. Previous

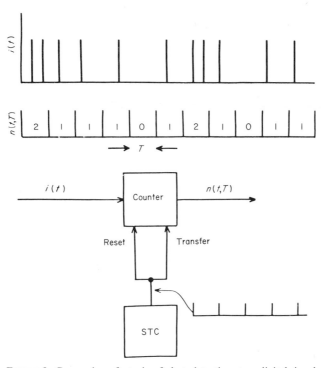

FIGURE 5. Conversion of a train of photodetections to a digital signal.

optical signals recorded with slower and less sensitive apparatus more often took the form of an analogue waveform made up from many time-smeared and overlapping detections.

How then does one process this type of signal? Figure 5 indicates the first step one might take. The train is converted to a series of integers by counting over consecutive time intervals generated by a sample time clock. The clock time is usually taken to be faster than the intrinsic fluctuations of $I(t)$ which one is trying to measure. One will then have a signal which is entirely digital in form say 2, 1, 1, 1 ... etc., recording the number of photo events which have occurred in each of these time intervals. The first sort of simple experiment (and I shall not say much here about this sort of experiment) that one can do is to look at the probability distribution. We have a mean; the mean in Fig. 5, for instance, is perhaps 1 or 1 and a bit events per interval: one can look at the fluctuations about that—sometimes one gets less and sometimes one gets more: the form of this distribution has been of interest for many years now and experiment dates from about 10 years ago. One did interesting experiments, for instance, looking at probability distributions and comparing them with theoretically predicted values for lasers, lamps and other sources. Figure 6 shows what happens when we take

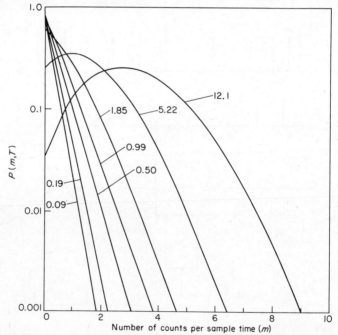

FIGURE 6. Photon counting distributions for a laser through threshold. The curves are labelled with the ratio of output power to the power at threshold.

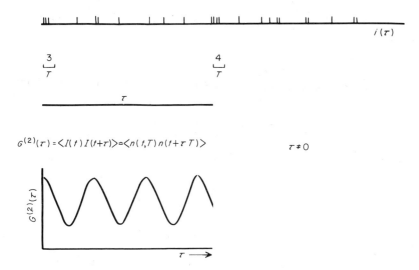

FIGURE 7. Photon correlation for two-mode spectrum.

a laser and detune it until it becomes a lamp. At the high output end we have a Poisson distribution, while at the opposite end it tends to a geometric distribution where the most probable value of intensity is zero.

I want to go beyond that sort of experiment, although one could talk for a long time about photon statistics as such, and get back to the extraction of spectral information from fluctuations, and in Fig. 7 I indicate what we have to do. Let us take two particular intervals out of the set, and here we have gone back to the two-mode spectrum so that there is a higher probability of a photodetection each period at the peak intensity values. At this particular time between a pair of samples there will be a positive correlation, and at the half period this correlation will drop.

We now define the intensity correlation function

$$G^{(2)}(\tau) = <I(0)\,I(\tau)>$$

We can show that this is the same, for non-zero delay, as the photon number correlation function

$$G^{(2)}(\tau) = <n(0, T)\,n(\tau, T)>$$

for sufficiently small T.

So here we have a function which tells us that there is a periodicity in the photon arrivals which is equal to the beat frequency that we are trying to observe. Clearly, if we wish to know what the frequency differences of the two lines are, all we have to do is look at the photon correlation function and measure the period.

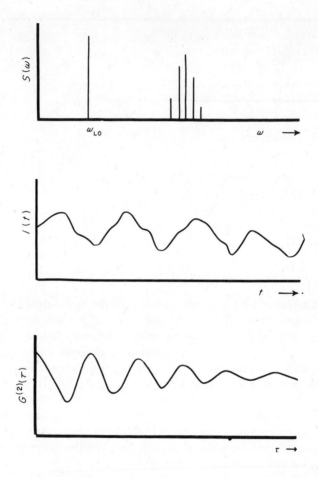

FIGURE 8. Photon correlation function for heterodyne case.

Now suppose we have a little broadening on one of those frequencies, (see Fig. 8). Then over a time short compared with the inverse mode spacing we get basically the original beat frequency. But the phase will decorrelate after a time equal to the inverse value of the broadened linewidth. That would be called a heterodyne experiment. We can move the local oscillator about and can put it in the middle of the broader spectrum and that we would call a homodyne experiment. What would happen here is that the basic beat frequency will get lower and lower, and we will eventually just see the decaying part of the correlation function and lose the oscillations altogether. In Fig. 9 we take away the local oscillator altogether, and we have a set of frequencies which are all beating with one another producing fluctuations which correlate with decay time equal to the inverse width of the spectrum folded with itself.

An important feature of these fluctuations is that if they arise from a laser scattering experiment, then the phase of the laser which is being scattered disappears from the expression for the correlation function for the intensity, because the intensity itself is $< \mathscr{E}\mathscr{E}^* >$ and is independent of the phase of $\mathscr{E}$, the exciting field. So we have the effect that was first known in the era of Hanbury Brown and Twiss and Forrester, Gudmundsen and Johnson: one is here doing a sort of spectroscopy where one is not limited at all by the phase stability of the laser. In fact, we recently had an enquiry from a Japanese laboratory—could we measure the Doppler shift of light scattering from a biological particle moving at the rate of a micron a minute? That, in terms of frequency, is about one cycle every sixty seconds. Clearly one couldn't do that if one was relying on the stability of the laser phase. Nevertheless, of course, the fluctuations still appear at the frequency one is looking for.

These are nice ideas in principle. One can extract spectral information

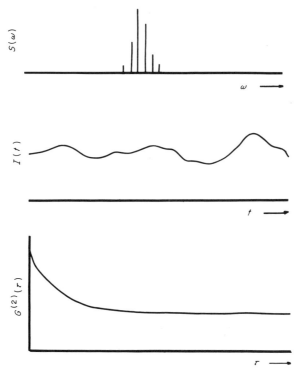

FIGURE 9. Photon correlation function of direct intensity fluctuations.

from intensity fluctuations; and the practical way one goes about it is to look at trains of photon numbers and correlate them with one another. The theory shows that for the heterodyne (homodyne) case, for any signal statistics,

$$g^{(2)}(\tau) = 1 + \frac{2P_s}{P_{LO}}(g^{(1)}(\tau)e^{i\omega_{LO}\tau} + c \cdot c)$$

where normalized correlation functions are defined

$$g^{(2)}(\tau) = <I(0)I(\tau)>/<I>^2$$
$$g^{(1)}(\tau) = <\hat{\mathscr{E}}^+(0)\hat{\mathscr{E}}^-(\tau)>/<I>$$

and P_s and P_{LO} are the signal and local oscillator powers respectively. The Fourier transform of the second-order correlation function in this case gives the optical spectrum

$$S^{(2)}(\omega) \sim S(\omega - \omega_{LO})$$

If no local oscillator is employed then the connection between optical spectra and intensity correlations can only be made for Gaussian statistics. In this case we have the Seigert relation (1943)

$$g^{(2)}(\tau) = 1 + |g^{(1)}(\tau)|^2$$

The Fourier transform gives the fold of the optical spectrum with itself

$$S^{(2)}(\omega) \approx \int S(\omega')S(\omega - \omega')d\omega'$$

We have developed an instrument to produce the photon correlation function $G^{(2)}(\tau)$. It is quite a different sort of spectrometer from the normal one, its input is just an electrical connection. This is plugged into a detector which is placed in the scattered light field. Of course, it doesn't have to be a scattered field—it could be a direct source and one could do the same measurements—but in principle as we will see in a moment, the narrow line experiments that we tend to want to do are from scattering experiments. There are, now, in fact, instruments of this nature either made by us or "home-made" in many laboratories around the world, doing experiments the nature of which are described in the last half of this paper.

Figure 10 is a chart of phenomena and techniques. I don't want to be controversial about Professor Smith's table, (this volume), but I have already mentioned that there is really no difficulty in looking at spectra of linewidths of about 10^6 times narrower than he was talking about! The chart shows just a selection of phenomena on which one might think about doing optical spectroscopy. The scale of frequencies starts on the right at optical frequencies of 10^{15} Hz or thereabouts. Resolutions of 1 part in 10^1, 10^2, 10^3 and 10^4 and so on are coming to the left, and one is happy with gratings down to about 10^{10} Hz. The Fabry-Perot is now routinely usable down to about 1 MHz; but there we really do run out of steam with the size of an interferometric instrument getting too large and too cumbersome to cope with. It is at this point that one turns over to the statistical methods of intensity correlation.

The sort of phenomena that one can look at in this lower range are Brownian motion—which I shall talk about, Rayleigh scattering just begins to come into it—that is of course entropy fluctuations, liquid crystal scattering extends down here obviously, fluid flow and turbulence I shall also talk about and also critical opalescence. From anything that moves and interacts with a light wave one can, if one thinks hard enough, extract information by observing the time scales of fluctuations that are going on and call it spectroscopy.

Normal spectroscopy in the grating region and to some extent in the Fabry-Perot region concerns very small objects. Nothing big can produce these sorts of frequencies—mechanical inertia will be too large. So in ordinary spectroscopy one is thinking of electrons, or even molecules, which are light

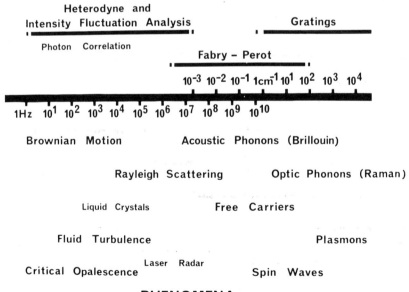

FIGURE 10. Chart of techniques and phenomena in optical scattering.

and fast. But there is no real reason to restrict one's consideration to such light and fast objects. One can begin to think now of spectroscopy of light fields which fluctuate on a much slower time scale and come right down to objects which are really macroscopic emitters.

We have heard a lot about Doppler shifts from moving atoms and molecules—we are now talking about Doppler shifts from moving macromolecules, or even tanks or aeroplanes!

In the last part of this paper I would like to describe a set of experiments which have been done using this sort of method. I think you will agree that there is a really staggeringly broad new area for optics that is being uncovered here. The plan of my presentation will be to start with small things moving and go on to bigger and bigger things moving.

Let us first of all talk about fluctuations which are of time scale at the top end of the range of the statistical techniques—fluctuations of entropy in a liquid, shall we say? In an ordinary pure liquid at a fixed temperature there are fluctuations of the dielectric constant which are due basically to two effects. In a simple liquid the entropy fluctuates at constant pressure, and the pressure fluctuates at constant entropy. The first is known as Rayleigh scattering or entropy fluctuation scattering and is controlled by the thermal conductivity of the medium. The second is known as Brillouin or first sound scattering, that is ordinary sound waves.

The sound-wave scattering can be very well studied with Fabry-Perot methods. The Rayleigh scattering runs on to the Fabry-Perot scale—just about—but it comes also on to the scale where one can look at it by intensity fluctuation methods and in Fig. 11 is shown the photon correlation function from light scattered from entropy fluctuations. This is an exponential with decay constant which goes as the square of the scattering vector times the thermal conductivity divided by the specific heat at constant pressure.

The correlation function is plotted on a logarithmic scale, so to refer back to Fig. 9, that is we have an exponentially decaying correlation of the light field with itself. The decay time here—11 microseconds—represents the decay time of the entropy fluctuations with the given wave vector which one can relate to the thermal diffusivity of the medium. This is for carbon tetrachloride, and it turns out—as it will in several other instances that I will describe—that this is now the best method available for measuring thermal diffusivity of liquids—by some order of magnitude over what was possible before with other non-optical methods. In fact, in order to confirm the value arrived at by this experiment, we looked up all the standard tables (one would think that the thermal conductivity of carbon tetracholoride was a known number). If one looks at all the standard tables, there are large differences in the predicted value, compared with the precision with which we are measuring here. The method, I believe, is now being used by the US

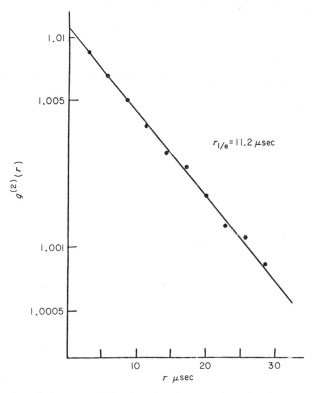

FIGURE 11. A typical autocorrelation function for scattering off carbon tetrachloride. The scattering angle was 3° 23′ and the temperature 20°C.

National Bureau of Standards.

Now the other prediction was that the slope or decay time of this function would change with angle as K^2, and this is shown at Fig. 12. Here we have the linewidth in kHz against the angle of scattering squared and it agrees with the expected K^2 dependence.

Moving on from pure liquids one can go to liquid mixtures, and Fig. 13 shows an acetone-carbon disulphide mixture. This is the same plot as the last figure of linewidth against K^2, showing in fact now a fluctuation which is due to the mutual diffusion coefficient of the one liquid in the other. So here we have a method for measuring mutual diffusion in two liquids, and one can see from the figure that it is also quite accurate.

One can move from the pure normal fluid, or the pure normal mixture, to the critical point of these materials, and now the entropy or concentration fluctuations become of longer range and slower. The result is that the spectrum of scattered light becomes narrower or, conversely, the photon

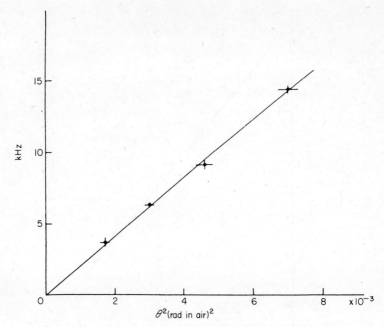

FIGURE 12. Angular dependence of Rayleigh linewidth for CCl_4.

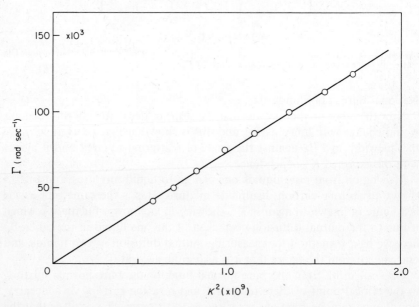

FIGURE 13. Correlation analysis of light scattered from a 10% V/V acetone/carbon disulphide mixture at 22°C.

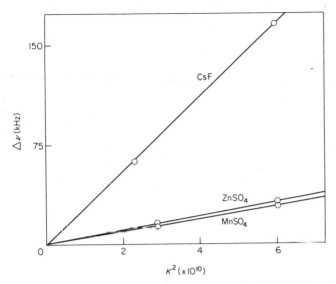

FIGURE 14. Correlation analysis of light scattered by various chemical solutions showing the diffusion-like angular dependence.

correlation function shows a longer decay time. This area of the study of critical phenomena by photon correlation methods is blossoming, and quite a number of laboratories are doing work in this field.

Going on to more macroscopic motions we ask what happens if we take, not a molecular motion like the entropy fluctuations or the Brillouin scattering, but actually put extra particles into a fluid in suspension. Figure 14 shows linewidths of the fluctuations extracted from such experiments. Here we have a correlation function which is basically an exponential of decay time given by the translation diffusion constant, which also has a K^2 dependence. One can relate this with spherical molecules to the actual radius of a particle by the Stokes-Einstein relation if one knows the particle is spherical. So here we have a method of actually measuring the radius of a particle or macromolecule. The smallest ones which we have measured are shown in the figure. These are just inorganic metal salts—zinc sulphate, magnesium sulphate, and caesium fluoride, again the linewidth against K^2 showing diffusive-like behaviour. Here we are actually measuring the size of the hydration shell around these ions in solution. This experiment has an interesting history. It was done because it was thought that the mono-molecular reaction that goes on in such an ionic solution should show up as a fluctuation in the light scattering, but unfortunately this effect is not large enough to be observable. Nevertheless, one can still do interesting experiments on small ions in solution.

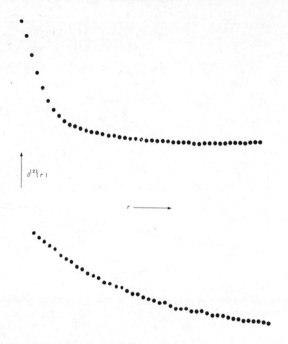

FIGURE 15. Vacancy diffusion in solid PMMA. The value of $g^{(2)}(0) - 1$ is about 0.01. The top curve has a time delay per channel of 1 ms and the bottom of 0.4 ms.

Figure 15 shows what happens if one looks at vacancies in a solid. This is a solid sample of polymethylmethacrylate, and here one sees two basic time scales, a two-component spectrum if you like, one of the vacancies diffusing connected with the motion of the side arms of the polymer and the other connected with the main chain of the polymer. These are very early studies: this is probably the first experiment of this sort in a solid, and the interpretation is still uncertain, but certainly there is an interesting field here in polymer physics. Also, polymers in solution are easier to study than solids, and there are a number of people working on that problem as well.

One can also study proteins or viruses, and Fig. 16 shows scattering from bovine serum albumin, the decay time gives the diffusion constant or the radius, and in this experiment we see the radius changing as we change the pH of the solution that the protein is in. The decay time gets longer as the molecule uncurls under the action of the changing pH.

Figure 17 shows scattering from a virus—adenovirus, giving a very accurate measurement of hydrodynamic radius. Again, in structural virology there is a lot of interest in this new technique for looking at sizes and hydrodynamic properties of viruses.

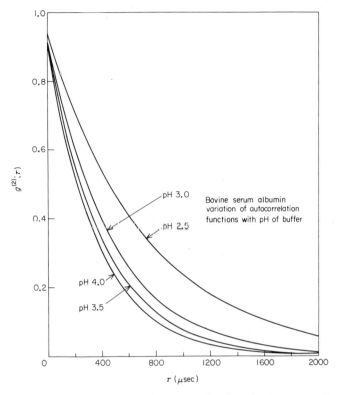

FIGURE 16. Photon correlation functions from scattering from bovine serum albumin as a function of buffer pH. The translational diffusion constants are approximately 5×10^{-7} cm^2/s.

Figure 18 shows scattering from polystyrene in toluene: that is a polymer solution and interestingly here I show the two methods—one with a homodyne reference beam, and one without, and we see that one gets a better statistical accuracy when one adds a reference beam, although it is not necessary and makes the experiment more complicated.

Figure 19 shows scattering from bull spermatozoa and we see, instead of the exponential functions that we have been having up till now, a more gaussian shape. For uniformly moving sperm the shape, theoretically, should be a zero-order spherical Bessel function. These are, if you like, Doppler shifts off individual spermatozoa, and so here we really see the velocities of sperm in sample, and one can measure spermatozoal motility this way. Again, there is a lot of interest for cattle breeders, and also for human studies, in the USA, this country and in France in this type of method.

Scattering from a retinal blood vessel at the back of the eye is shown in Fig. 20. Here one is measuring the Doppler shift off the red corpuscles, so one

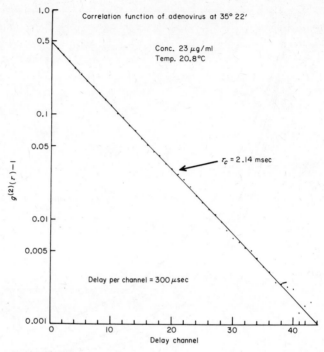

FIGURE 17. Photon correlation function from scattering from adenovirus. The translational diffusion constant is approximately 5×10^{-8} cm^2/s.

can measure the blood flow velocity. This is of interest for ophthalmologists in diagnosing certain types of retinal disease—again there is work at MIT, where the technique originated in the group of Professor Benedek, and in this country at the Institute of Ophthalmology and the Royal College of Surgeons.

The last type of experiment I want to discuss is when we really do get large particles moving about; these occur in situations where we wish to study a flowing fluid. The Doppler shifts from very fast-moving suspended particles can still be rather large and, in order to reduce them, we tend to use a method known as the differential Doppler technique. In this we take two laser beams; then in a given scattering direction, from a particle moving at a given velocity, we get two Doppler frequencies and we can measure the difference. This is known as the difference Doppler signal, and comes down nicely on to the time scale of fluctuations which we can observe conveniently, and of course we have some control in the angle between the two beams. It does have the further advantage that the actual direction of the scattered beam is not involved in the expression for the difference frequency, so there is no k-vector dependence and a very large collecting aperture can be used.

There are many applications of this type of technique, a whole new field in flow measurement has grown up. One has now whole conferences on the subject—known as laser Doppler anemometry, or laser Doppler velocimetry. Let me just show one or two results. Figure 21 shows a setup used at the Royal Aircraft Establishment at Farnborough for looking at velocities of air

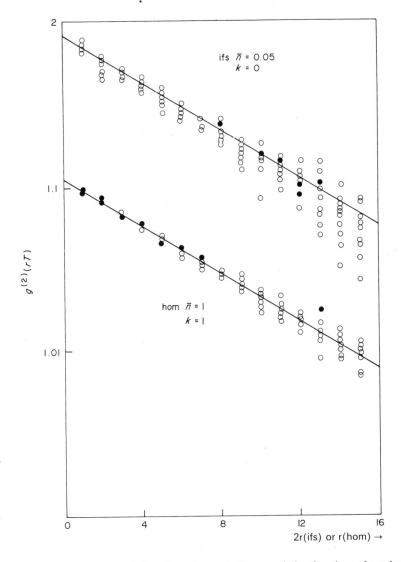

FIGURE 18. Homodyne and direct intensity fluctuation correlation functions of a polystyrene fraction in toluene.

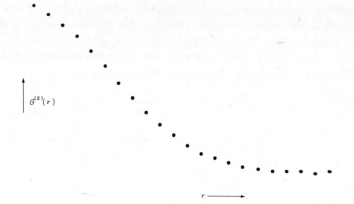

FIGURE 19. Photon correlation function from scattering from bull semen. The time delay per channel is 1 ms representing a mean sperm velocity of approximately 100 μm/s.

flowing in a supersonic wind tunnel. An argon-ion laser beam is split to produce two beams in each of two colours which allow measurements of velocity in two directions. The beauty of this method of looking at velocities in a wind tunnel is that one does not have to put anything inside, and one gets a new range of experiments possible where interference with the flow is quite negligible.

Figure 22 gives results obtained around a cone in supersonic flow. Here we have velocity against distance from the cone surface to the shock wave. The expected value due to the theory agrees with the experimental results that are obtained. I should mention that these rather nice experiments have been done by Mr J. Abbiss of the Royal Aircraft Establishment.

The last two figures show what happens when one does not a single velocity in the flow but several. We have a narrow spectrum, where one does not get a single frequency due to the Doppler difference spectrum but a decaying correlation function as in Fig. 8 and the decay here can be translated into a

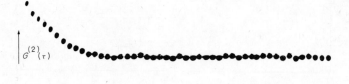

FIGURE 20. Photon correlation function from scattering from a retinal blood vessel. The time delay per channel is 2 μs. Typical flow rates in a vessel of 100 μm diameter are of the order of 1 cm/s. Exposure was 100 msec with 10 μW of the He Ne laser light.

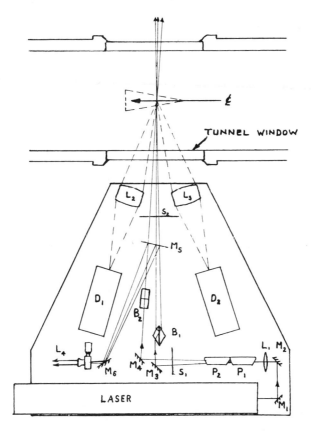

FIGURE 21. Two dimensional Doppler anemometer system for use in supersonic wind tunnel flows.

degree of turbulence of the flow. We thus have a direct method of measuring not only the velocity of a flow but also its turbulence intensity and this is often just as useful as the actual velocity measurement. Figure 23 shows a correlation function from a turbulent open air jet. Work of this kind is being done by the British Gas Board in connection with the design of gas burners, optimizing mixing in these turbulent situations.

Figure 24 shows an application to noise-reduction studies in the Concorde engine—the Olympus 593. Here one has a measurement of mean velocity versus revolutions per minute at a particular point in the flow behind the engine, together with the intensity of the turbulence in that flow. One can put various bits inside the engine and see what happens to the turbulence levels—the turbulence, of course, being responsible for noise production. The RB211 engine has also been studied in this way.

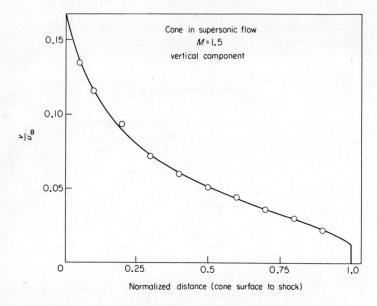

FIGURE 22. Supersonic flow around a 10° cone.

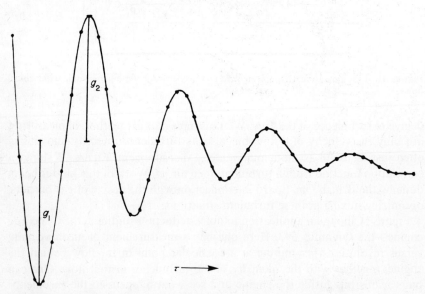

FIGURE 23. Photon correlation function from scattering from particles following a turbulent air flow.

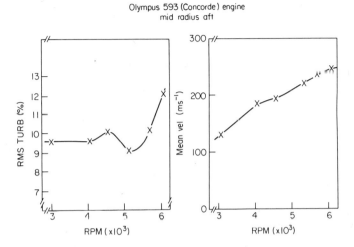

FIGURE 24. Flow and turbulence measurements on the Olympus 593 aero engine.

To conclude I hope that I have been able to give an indication of the very wide range of applications which are arising from this new area of high resolution spectroscopy by photon correlation methods and to convey the basic ideas lying behind them. I am quite sure that this is a field which will enjoy further expansion and consolidation in the years to come and I would not at this time even like to predict in which area its maximum impact will lie.

I have not given explicit references in this review as we should have had a very large number. All the figures, with the exceptions of 21 and 22 from RAE and 24 which is printed with the kind permission of Rolls-Royce, Derby, show results obtained by my colleagues and me at RRE Malvern. If further reading and an introduction to the literature is required, many of the basic ideas and applications can be found in the proceedings of a recent NATO Advanced Study Institute "Photon Correlation and Light Beating Spectroscopy" eds. H. Z. Cummins and E. R. Pike, Plenum Press 1974.

Chapter 4

High Resolution Tunable Infrared Lasers*

by A. Mooradian, Lincoln Laboratory, Massachusetts Institute of Technology, Lexington, Massachusetts 02173, U.S.A.

This paper discusses some of the technology of tunable lasers which operate in the near to middle infrared regions of the spectrum. Included among these are semiconductor diode lasers, spin-flip Raman lasers, parametric oscillators, and nonlinear mixing using tunable and quasitunable laser sources. The spectral linewidth and tuning range of these devices will be discussed, with emphasis on their application for high resolution spectroscopy.

SEMICONDUCTOR DIODE LASERS

Semiconductor diode lasers[1] have been used for high resolution spectroscopic studies of gases in the middle and near infrared more extensively than any other presently available tunable infrared laser. The wavelength ranges covered by compound semiconductors in which diode lasers have been fabricated to date are shown in Fig. 1. By choosing the appropriate composition of one of these compounds, lasers can be produced at any desired wavelength in the broad range from 0.6 to 32 μm. An individual diode laser can be tuned by a magnetic field, hydrostatic pressure, or by temperature. A specific operating temperature is usually chosen using a fixed temperature bath of liquid helium or liquid nitrogen, although variable temperature refrigerators presently have sufficient temperature stability for most applications. Most diode lasers have a spectral output consisting of several longitu-

---

*This work was sponsored by the Department of the Air Force, NSF/RANN, and the Atomic Energy Commission under a subcontract from the Los Alamos Scientific Laboratory.

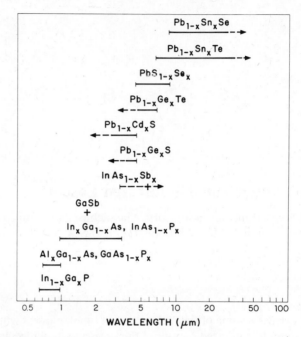

FIGURE 1. Wavelength ranges covered by various semiconductor diode laser materials. The solid lines represent ranges covered at present. The dotted portions and the arrows indicate regions where lasers may be developed in the future.

dinal and transverse modes which increase in number and shift in frequency with increased drive current. The fine tuning of each of these modes with injection current is shown[2] in Fig. 2. Heating of the diode due to nonradiative as well as I^2R losses causes a change in refractive index of the mode volume which changes the effective cavity length. Continuous tuning of a single mode from this effect can be several cm^{-1} before a mode jump occurs. The effect of frequency tuning due to a change in cavity length from thermal expansion is negligible by comparison. The jumping of modes results from a shift of peak frequency of the gain spectrum as the energy gap of the material changes. Even though this type of discontinuous tuning occurs in all diodes, complete wavelength coverage can be achieved when a number of adjacent modes overlap. One of the modes lying close to the absorption line of interest is selected by a spectrometer and then fine tuned by changing the current through the diode. The continuous tuning range of a single mode from a diode laser is quite adequate for scanning a pressure broadened gas absorption line in the infrared which is typically about 0.1 cm^{-1} (3 GHz) wide. Figure 3 shows the absorption spectrum of H_2O obtained by current-tuning a $PbS_{1-x}Se_x$ diode[3] laser, whose emission was passed through a cell

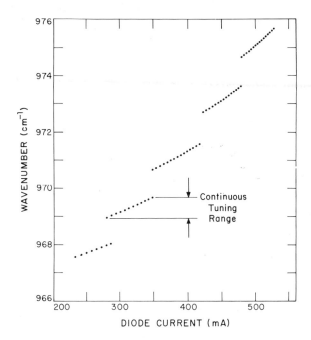

FIGURE 2. Fine tuning of modes from a $Pb_xSn_{1-x}Te$ diode laser as a function of injection current. Tuning is continuous within a mode. (From Ref. 2)

containing H_2O at a reduced pressure, as well as atmospheric H_2O in the laboratory. The data were obtained by slowly sweeping the diode laser current, while mechanically chopping the laser emission and synchronously detecting the transmitted radiation. The relative frequency shift as a function of diode current was determined by using a germanium etalon. The transmission through the etalon is shown in the lower portion of Fig. 3. The difference between the pressure-broadened and Doppler-broadened linewidths is clearly seen as well as the pressure shift of the centre frequency.

The output frequency of a laser in the case of a homogeneously broadened Lorentzian line is given by

$$v = \frac{v_c\Gamma_s + v_s\Gamma_c}{\Gamma_c + \Gamma_s} \quad (1)$$

where v_s is the peak of the gain frequency, Γ_s the linewidth (full width at half height) of the gain, v_c and Γ_c the corresponding cavity parameters. The cavity linewidth measured in cm^{-1} is given by

$$\Gamma_c = \left[\alpha - \frac{\ell n R_1 R_2}{2\ell}\right] \bigg/ \left(n + v\frac{\partial n}{\partial v}\right) \quad (2)$$

where n is the refractive index, α is an effective loss parameter in cm^{-1}, R_1 and R_2 are the end mirror reflectivities, and ℓ is the cavity length in cm. In the case of semiconductor lasers, the spontaneous linewidth, which is not necessarily Lorentzian, is much greater than the cavity linewidth and the output frequency is primarily given by the temperature-dependent cavity frequency

$$v_c = cM/2\ell\left(n + v\frac{\partial n}{\partial v}\right) \qquad (3)$$

where M is a large integer and

$$n \approx n_0 + (dn/dT)\Delta T. \qquad (4)$$

Cavity tuning rates for typical lead-salt semiconductor lasers are usually

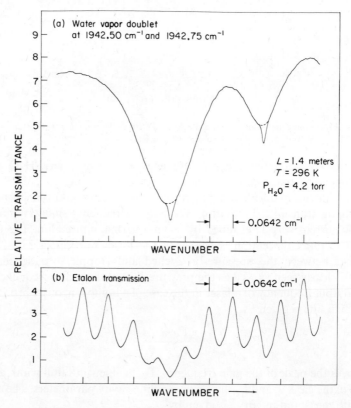

FIGURE 3. (a) Relative transmittance of the $6_{5,2} \leftarrow 5_{4,1}$ and $6_{5,1} \leftarrow 5_{4,2}$ doublet near 1942.5 cm^{-1} showing the negative pressure shift of atmospheric pressure air broadened lines from the narrow low-pressure lines. (b) The transmission of a germanium etalon as a function of laser frequency near 1942.5 cm^{-1}. (From Ref. 3)

about one-third the bandgap tuning rates with temperature. Increasing the drive current can not only shift the gain peak but also increase its magnitude and change its lineshape, which would complicate any detailed mode tuning analysis.

Most semiconductor diode lasers[1] are presently fabricated in a "stripe geometry" in order to keep the laser output in the fundamental transverse mode. Multilongitudinal mode output occurs in these devices, however, at drive currents significantly above threshold. Initial studies of the longitudinal mode control of a semiconductor laser have been made with the use of a grating controlled cavity GaAs[4-6] laser in order to improve the usefulness of the spectral output of these devices. The spectral properties of both a 77 K continuously operating[5] and a room temperature[4, 6] pulsed external cavity grating-controlled GaAs diode laser have been studied. More recently,

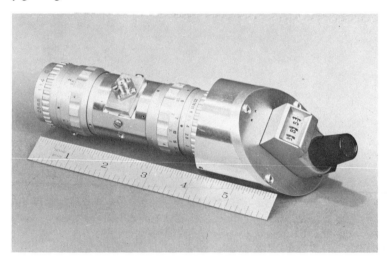

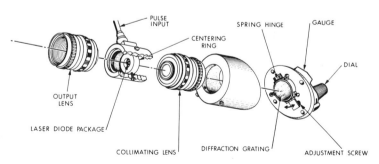

FIGURE 4. Photograph of a room-temperature external-cavity gallium arsenide diode laser (top). Schematic of device (bottom). (From Ref. 7)

[6] a grating-controlled room temperature GaAs diode laser has operated with a spectral width of less than 0.2 Å and a peak output power of nearly 3 W. It is interesting to note that the same diode operating with a 100% reflector in place of the grating produced 3.5 W in a bandwidth of over 40 Å. Figure 4 shows a photograph of this device. Only one end of the diode is antireflection coated. Figure 5 shows the output power as a function of wavelength for a number of GaAs external cavity grating-controlled diode lasers at room temperature.[7] Tuning has been achieved from 8300 to 9300 Å with

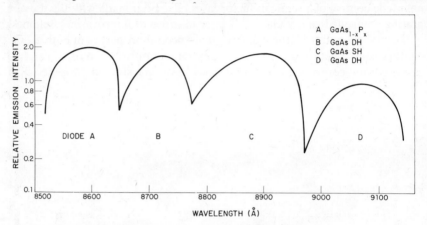

FIGURE 5. Peak output power as a function of wavelength for a $GaAs_xP_{1-x}$ and three different GaAs external cavity grating-controlled diode lasers at room temperature. Different amounts of silicon doping extend the GaAs laser wavelengths. (From Ref. 7)

a spectral width of less than 0.4 Å. For temperatures below 77°K, output powers of up to 1 W cw in a single frequency should be possible from GaAs. A diode laser of $Pb_xSn_{1-x}Te$ has operated with an output near[8] 10 μm using an external cavity consisting of a single concave mirror. This device operated near liquid helium temperature with only one end of the diode antireflection coated.

The spectral output of these grating controlled semiconductor diode lasers still occurs in a number of axial modes for pumping levels much above threshold. Additional intercavity mode control elements would be necessary to limit the output to a single mode. For a number of spectroscopic applications, single frequency operation may not be necessary or desired. An appropriate external cavity device operated pulsed could have its modes chirped enough to fill in the spectral envelope defined by the external cavity grating dispersion and the mode width defined by the junction. Such an envelope width could be significantly less than 0.1 cm^{-1} in width if the diode were not driven very far above threshold. Tuning would effectively be continuous for spectroscopic lines that were broader than the envelope width.

The application of hydrostatic pressure to a semiconductor laser can provide a very broad tuning range for a single device. Figure 6 shows bandgap tuning characteristics for a number of binary lead-salt and III-V compound semiconductors at 77 K. A lead selenide diode laser at 77 K has been tuned[9] from 7.5 to 22 μm using hydrostatic pressure up to 14 kilobars. The stability of the hydrostatic pressure has been controlled sufficiently to perform Doppler-limited spectroscopy using a pressure-tuned laser of GaAs[10] and PbS_xSe_{1-x}. Uniaxial tuning of diode lasers can avoid the limitations of low temperature operation with helium as the pressure transmitting fluid; however, in practice it is difficult to avoid crushing the lasers. Uniaxial pressure tuning rates are about one-third the hydrostatic rates.

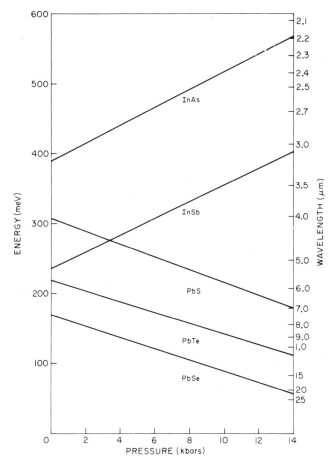

FIGURE 6. Bandgap variation with hydrostatic pressure for some binary semiconductor compounds at 77 K. (From Ref. 10)

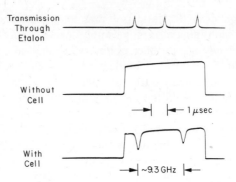

FIGURE 7. Absorption spectrum of the Doppler limited $6s^2S_{1/2} \to 6p2P_{3/2}$ transition in caesium using a 77 K chirped GaAs diode laser ($\lambda = 8521.1$ Å, hydrostatic pressure = 160 bars). Top: transmission of chirped axial mode of GaAs diode through the free spectral range of a Fabry-Perot interferometer. Middle: mode intensity without absorption cell. Bottom: mode intensity with caesium cell. (From Ref. 10)

The semiconductor laser is coarsely tuned to the absorption lines of interest by setting the hydrostatic pressure, and the fine tuning is done by varying the temperature of the junction with dc current in the case of cw operation, or by the frequency chirp which occurs in pulsed operation. Figure 7 shows an example[10] of the latter, where a GaAs diode laser is used to look at the real time, Doppler-limited absorption spectrum of the $6s^2S1/2 \to 6p^2P3/2$ transition in caesium. Since helium freezes at these high pressures for temperatures less than 77 K, most of the diodes to date have been operated at 77 K in the pulsed mode with chirp rates low enough to sweep through a Doppler linewidth in a time longer than the response time of the detector. The recent development[11] of infrared diodes which operate continuously at 77K or higher will greatly facilitate the usefulness of hydrostatic pressure tuning. Because of the broad-band wavelength coverage using hydrostatic pressure tuning, only the binary semiconductor compounds and one or two alloy semiconductors would be necessary to cover the wavelength range from 2 to 35 μm.

Spin-Flip Raman Lasers

Spin-flip Raman lasers have been developed over the past few years to the point of being useful for a number of spectroscopic applications. The schematic diagram for a typical spin-flip laser system is shown in Fig. 8. Molecular gas lasers such as CO, CO_2, and HF have been used to pump semiconductor crystals such as n-type InSb, InAs or HgCdTe in a magnetic field at low temperatures. The output characteristics of these devices are listed in Table 1.

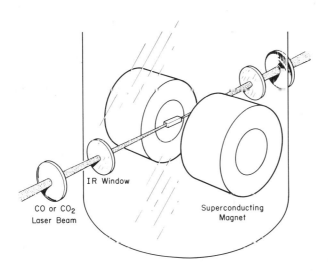

SPIN-FLIP LASER

FIGURE 8. Schematic diagram of spin-flip Raman laser. Magnetic field direction is normal to the direction of the pump laser beam.

TABLE 1. Spin-flip laser characteristics

Pump Laser	Material	Wavelength Range	Maximum Output in Range	Ref.
pulsed CO_2 10.6 μm	InSb	9–14.6 μm	1 kW in 1st Stokes 10 W in 2nd Stokes 30 W in 1st anti-Stokes	20
SH of pulsed CO_2 5.3 μm	InSb	5.2–6 μm	~ 1 kW	21
CW CO 5.2–6.5 μm	InSb	5.2–6.5 μm	1 W CW in 1st Stokes comp.	22
pulsed HF 2.9 μm	InAs	2.98–3.00 μm	> 200 W in 1st Stokes	18
pulsed CO_2 9.52 μm	HgCdTe	9.7–10.2 μm	1 W pulsed in 1st Stokes	19
pulsed dye laser	CdS	~4925 Å		23

The output linewidth and frequency tuning characteristics are particularly important for use in high resolution spectroscopic applications. As in the case of semiconductor diode lasers, the mode tuning properties are governed by Equations (1)–(4). The spontaneous spin-flip Raman linewidth is one of the important parameters which determines not only the laser threshold but the mode-tuning characteristics. Recently,[12] linewidths as small as 200 MHz have been observed using a small-signal gain technique. A more detailed experimental and theoretical analysis of this lineshape as a function of magnetic field[13] and temperature[14] has been reported, also using this technique. A sample of InSb was pumped with a CO laser at frequency ω_1 below threshold but with enough gain over the sample length to produce about a 10% amplification at frequency $\omega_2 = \omega_1 = g\beta H$. The intensity of a weak colinear probe beam at ω_2 was measured as the magnetic field was swept. The linewidth, which is very close to Lorentzian, increased from 200 MHz to 3000 MHz with magnetic field at least up to 10 kG in low-concentration samples. The tuning rate of the mode[15] frequency with magnetic field has been measured to vary from 16 MHz/G[12] for a sample with $n = 1 \times 10^{16}$ cm^{-3} to greater than 60 MHz/G[14] for samples with $n \gtrsim 10^{15}$ cm^{-3} at low magnetic fields, which approaches the 67.5 MHz/G tuning rate of the spontaneous peak observed in the same field range. Figure 9

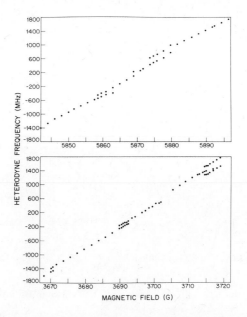

FIGURE 9. Single-mode tuning characteristics of a cw spin-flip laser ($n_e = 8 \times 10^{14}$ cm^{-3}). (From Ref. 15)

shows a typical measured tuning rate of a single mode from a low-field spin-flip laser taken using a heterodyning technique. The tuning rate within a mode depends in part upon the electron concentration in the sample, magnetic field, temperature, and cavity parameters. Shorter cavities could provide not only a larger longitudinal mode separation, but a greater range of continuous mode tuning before a jump occurred. In practice, care must be taken to insure that the spin-flip output occurs in a single frequency mode when use for high resolution spectroscopy is desired. Depending upon the alignment between the pump beam and the laser crystal as well as a number of other parameters, the spectral output of the spin-flip laser can vary from a nearly continuously tunable broad-band (100 MHz) output to a quasi-continuous output in which one or more modes tune over a small frequency range before a mode jump occurs.

Recently, a hybrid frequency and phase lock closed loop system, to maintain the output frequency of a spin-flip laser at a variable frequency offset from the frequency of a CO local oscillator laser, has been constructed.[15] An error signal derived by heterodyning the spin-flip output with a stable CO laser together with a stable microwave oscillator in a HgCdTe photodetector has been used to drive small modulation coils around the sample. The output heterodyne signal of a frequency stabilized spin-flip laser mixed with a stable CO laser is shown in Fig. 10. Precision frequency control is achieved by tuning the microwave oscillator. A usable linewidth and long and short term frequency stability of better than 30 kHz (limited by the stability of the two CO lasers) have been achieved. This technique also provides absolute frequency tuning and calibration against CO laser secondary frequency standards. Because beat frequencies in excess of 60 GHz have been directly measured using HgCdTe[16] photo-

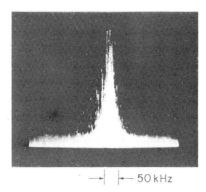

FIGURE 10. Photograph of spectrum analyser trace showing the spectral output of a "locked" cw spin-flip laser (bandwidth 3 kHz, 10 second exposure; $n_e = 8 \times 10^{14}$ cm^{-3}, $H = 3760$ G). (From Ref. 15)

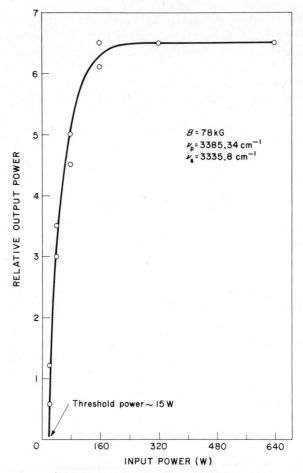

FIGURE 11. Relative InAs spin-flip laser output power as a function of the input HF laser pump power ($n_e \sim 1.4 \times 10^{16}$ cm^{-3}, $H = 78$ kG, and $T = 2$ K). (From Ref. 18)

diodes, complete coverage between the rotational lines of CO and CO_2 lasers is possible. Absolute frequency rather than wavelength can now be measured with this technique, almost anywhere in the infrared using nonlinear mixing between a CO and a CO_2 laser in an appropriate nonlinear crystal.

A free-running spin-flip laser has been recently[17] used to observe the Lamb dip absorption of water vapour in the 5 μm region. Natural linewidths as small as 100 kHz were measured.

Stimulated spin-flip Raman scattering in InAs has been[18] achieved using an HF–TEA laser (2 → 1 band, P(9) line at 3385.34 cm^{-1}) to pump near the bandgap resonance. Tunable emission was observed from 3347 to 3332 cm^{-1}.

The output power in the first Stokes component increased monotonically with magnetic field and conversion efficiencies in excess of ten percent were observed. Figure 11 shows the relative output power as a function of the input power for the InAs spin-flip laser. At the highest pump intensities, the spin-flip output pulse as measured on an oscilloscope became flat-topped and did not increase with pump intensity, indicating the onset of spin saturations. Crystals with lower electron concentrations are necessary in order to lower the pump intensity as well as the magnetic fields necessary to reach threshold.

Spin-flip Raman scattering which was interpreted as stimulated has also been reported[19] using a CO_2–TEA laser resonantly to pump an n-type crystal of $Hg_xCd_{1-x}Te$. High quality crystals must be developed in order to make this a useful spectroscopic device.

Nonlinear Mixing

One of the potentially useful ways of producing tunable coherent IR radiation over a broad band is sum and difference frequency generation in a nonlinear material using a fixed and a tunable frequency source. There have been a number of successful experiments in which a dye laser plus either a ruby or a doubled Nd:YAG laser have been used to generate tunable radiation in the UV–visible–near IR using $LiNbO_3$ as the nonlinear crystal. More recently,[24] a stabilized cw dye laser and a cw single frequency argon-ion laser have been mixed in $LiNbO_3$ to generate a cw tunable output from 2.2–4.2 μm with sufficient power to do linear absorption spectroscopy. The infrared frequency stability was about 20 MHz which is adequate for Doppler limited spectroscopy in this region. Output power exceeded one μW for 10 mW and 100 mW of input dye and argon laser power, respectively. Figure 12 shows the experimental arrangement for the difference-fre-

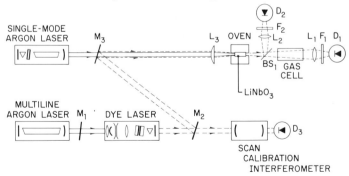

FIGURE 12. Experimental arrangement for difference frequency generation in $LiNbO_3$ using a frequency stabilized argon ion laser and a tunable, stable cw dye laser. (From Ref. 24)

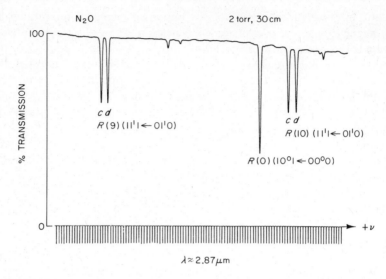

FIGURE 13. Experimental trace of the absorption spectrum of N_2O taken using the difference-frequency spectrometer of Fig. 12. Frequency marker lines are 150 MHz apart and are generated using a fixed Fabry–Perot etalon to monitor the tuning of the dye laser. (From Ref. 24)

quency spectrometer. This system has routinely recorded spectra such as that shown in Fig. 13. Some of the difficulties associated with using lasers in the visible to generate IR is that nonlinear materials such as $LiNbO_3$ do not stand up well to intense visible–UV radiation; many nonlinear materials which are highly transparent in the visible do not transmit much beyond 5 µm (new crystals such as $AgGaS_2$ may eliminate this problem). The photon energy loss also limits the maximum power conversion efficiency.

An alternative approach to generating stable-frequency tunable radiation in the infrared is the use of infrared nonlinear materials such as the chalcopyrites. Some of the advantages of these materials include good IR transmission properties and high nonlinear figures-of-merit. Recently,[25] a cw CO and CO_2 laser have been mixed in $CdGeAs_2$ to generate quasi-tunable infrared radiation in the range from 2.5–17 µm. Using $CdGeAs_2$, internal conversion efficiencies of 25% have been measured for second harmonic generation using a pulsed 10 µm CO_2 laser. For difference-frequency generation, an output power at 12.87 µm, for example, of 4 µW has been measured from $CdGeAs_2$ using 97 and 1250 mW cw of CO and CO_2 laser radiation, respectively. Figure 14 shows the phase-matching angle for various wavelengths which have been obtained in $CdGeAs_2$. The use of a high pressure gas laser such as CO_2 which has its rotational lines sufficiently broadened to have complete overlap can provide a broad-band tunable

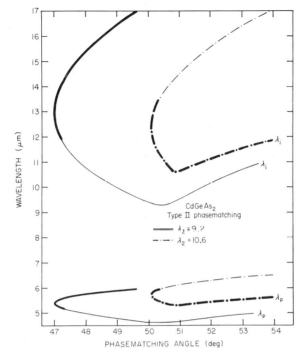

FIGURE 14. Phase-matching angle *versus* wavelength for $CdGeAs_2$ for difference-frequency generation. Heavier curves are experimentally obtained data. (From Ref. 25)

laser source over much of the infrared using nonlinear mixing in crystals such as $CdGeAs_2$.

Optical parametric oscillators[26,27] have also been used not only as the tunable laser source, but to down-convert by difference-frequency generation using a nonlinear crystal. Such systems have the advantage of room-temperature operation and broad-band wavelength coverage. Because of the inherently large gain bandwidth in parametric oscillators, practical devices with linewidths less than 0.1 cm^{-1} have been difficult to achieve.

REFERENCES

1. For a more comprehensive review see "Tunable Semiconductor Diode Lasers and Applications", Melngailis, I., and Mooradian, A. (1975), *in* "Laser Applications to Optics and Spectroscopy", Sargent, M., and Jacobs, S. F. eds., pp. 1–51, Addison-Wesley, Reading, Massachusetts.
2. Hinkley, E. D., Blum, F. A., and Nill, K. W., *in* "Laser Spectroscopy of Atoms and Molecules", Walther, H. ed., a volume in the series Current Topics in Physics, Springer-Verlag, Heidelberg (to be published).
3. Eng, R. S., Kelley, P. L., Calawa, A. R., Harman, T. C., and Nill, K. W. (1974), *Molec. Phys.*, **28**, 653–664.

4. Edmunds, H. D., and Smith, A. W. (1970), *IEEE J. Quantum Electron.*, **6,** 356.
5. Ludeke, R. and Harris, E. P. (1972), *Appl. Phys. Lett.*, **20,** 499.
6. Rossi, J. A., Chinn, S. R., and Heckscher, H. (1973), *Appl. Phys. Lett.*, **23,** 25.
7. Heckscher, M., and Rossi, J. A. (1975), *Appl. Optics*, **14,** 94.
8. Johnson, E. J., private communication.
9. Besson, J. M., Butler, J. F., Calawa, A. R., Paul, W., and Rediker, R. H. (1965), *Appl. Phys. Lett.*, **7,** 206.
10. Pine, A. S., Glassbrenner, C. J., and Kafalas, J. A. (1973), *IEEE J. Quantum Electron.*, **9,** 800.
11. Groves, S. H., Nill, K. W., and Strauss, A. J. (1974), *Appl. Phys. Lett.*, **25,** 331.
12. Mooradian, A. (1972), *in* "Fundamental and Applied Laser Physics", *Proceedings of the Esfahan Symposium, 1971*, Feld, M. S., Javan, A., and Kurnit, N. eds., pp. 613–627 addendum, John Wiley and Sons.
13. Brueck, S. R. J., and Mooradian, A. (1973), *Optics Communs.*, **8,** 263–266.
14. De Silets, C. S., and Patel, C. K. N. (1973), *Appl. Phys. Lett.*, **22,** 543–545
15. Brueck, S. R. J., and Mooradian, A. (1974), *IEEE J. Quantum Electron.*, **10,** 634.
16. Spears, D. L., and Freed, C. (1973), *Appl. Phys. Lett.*, **23,** 445–447.
17. Patel, C. K. N. (1974), *Appl. Phys. Lett.*, **25,** 112.
18. Eng, R. S., Mooradian, A., and Fetterman, H. R. (1974), *Appl. Phys. Lett.*, **25,** 453.
19. Sattler, J. P., Weber, B. A., and Nemarich, J. (1974), *Appl. Phys. Lett.*, **25,** 491.
20. Aggarwal, R. L., Lax, B., Chase, C. E., Pidgeon, C. R. and Limbert, D. (1971), *Appl. Phys. Lett.*, **18,** 383.
21. Wood, R. A., McNeish, A., Pidgeon, C. R., and Smith, S. D. (1973), *J. Phys. Chem.*, **6,** L144.
22. Brueck, S. R. J., and Mooradian, A. (1971), *Appl. Phys. Lett.*, **18,** 229.
23. Scott, J. F., and Damen, T. C. (1972), *Phys. Rev. Lett.*, **29,** 107.
24. Pine, A. S. (1974), *J. Opt. Soc. Am.*, **64,** 1683.
25. Kildal, H., and Mikkelsen, J. C. (1974), *Optics Communs*, **10,** 306.
26. Byer, R. L. (1974), *in* "Laser Spectroscopy", Brewer, R. G. and Mooradian, A. eds., Proceedings of a Conference held in Vail, Colorado, 25–29 June, 1973, Plenum Press.
27. Hanna, D. C., Luther-Davies, B., Smith, R. C., and Wyatt, R. (1974), *Appl. Phys. Lett.*, **25,** 142.

Chapter 5

Generation and Measurement of Ultra-short Pulses
by D. J. Bradley, Imperial College, London SW7, England

As Professor Porter has pointed out, the relationship between pulse duration and bandwidth arises, simply, from the uncertainty principle. Perhaps I should point out initially that the generation of ultra-short light pulses by lasers was for many years an oddity and not taken too seriously. However it is now clear that this aspect of laser technology is going to be very significant. Already there are two important fields of application. The first is the use of high energy, short pulses for the compression of matter to very high densities and temperatures for thermonuclear fusion[1] and there are large laboratories being assembled for this in the USA and USSR. Without the earlier work on the generation and measurement of short pulses it just would not have been possible to consider this as a practical proposition. The other important field of application is to photochemistry and photobiology.

What I would like to do is to discuss some of the fundamental processes in the generation and measurement of ultra-short pulses. Inevitably, a lot of this will be well known to many here, but I thought that it might be useful to summarize the general principles. Then I would like to go very quickly to the current state of the art. The origin of the picosecond pulses (10^{-12} sec) which are now obtainable from neodymium: glass lasers and from dye lasers is, of course, the very broad-band fluorescence of these laser media, which will carry a pulse of duration about one tenth of a picosecond. As Dr. Pike has pointed out, an examination with sufficient time resolution of the intensity of any light source, with a bandwidth of that extent, would

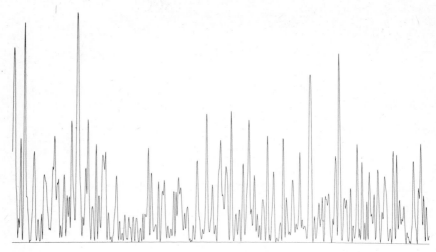

FIGURE 1. Intensity distribution of broad-band light source. For a bandwidth of $\sim 100 \, \text{cm}^{-1}$, duration of shortest fluctuation spike is 10^{-13} sec.

show the kind of intensity distribution shown in Fig. 1.

How does the generation of short pulses proceed inside a laser cavity? Unfortunately the term "mode-locking" as normally applied tends to obscure the simplicity of the phenomenon. The broad-band fluorescence from which laser action builds up has a similar intensity fluctuation pattern to that shown in Fig. 1. The trick is to select one of the spikes out of the initial intensity distribution and to amplify it preferentially so that eventually all the laser energy is concentrated in a single pulse of duration of a few picoseconds. This mechanism of selection and amplification of a fluctuation spike is easily achieved by the simple expedient of inserting into the laser resonator a saturable absorber dye. Since organic dyes play a dominant role in picosecond pulse generation as well as being efficient laser media we should take a closer look at them. Figure 2 shows the molecular structures of the families of dyes commonly used in dye lasers.[2] These are large polyatomic molecules, planar in form and containing conjugated bonds. As regards the interaction of light, it is only necessary to consider the two π-electron clouds, above and below the plane of the molecule. Figure 3 gives the essential molecular parameters which we need to consider in describing the generation of ultra-short pulses. There is a series of singlet and triplet levels. The pertinent points to remember are the very-broad-band absorption, which is typically hundreds of wavenumbers wide, and the mirror symmetry of the corresponding fluorescence band which is Stokes-shifted to longer wavelengths. Relaxation among the rotational and vibrational levels of the electronic states occurs very rapidly because each large molecule undergoes

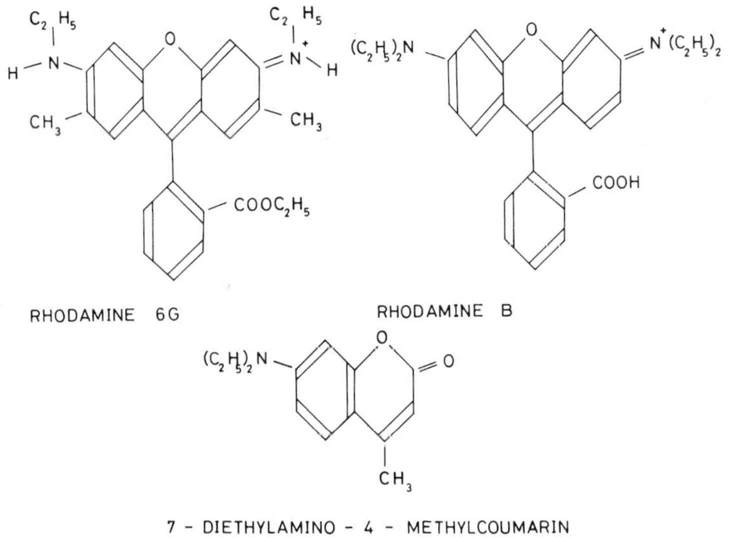

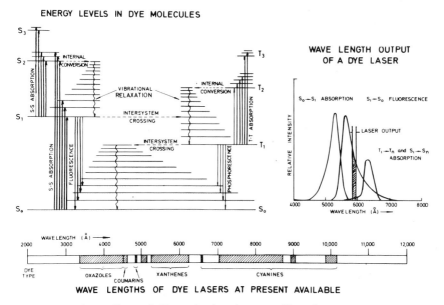

FIGURE 2. Structures of dye molecules employed as laser active media.

FIGURE 3. Energy levels and spectra of laser dyes.

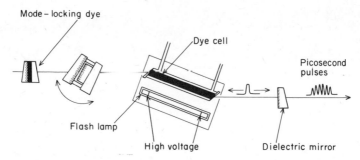

FIGURE 4. Passively mode-locked dye laser cavity arrangements.

at least 10^{12} collisions per second with the surrounding solvent molecules. Equilibrium is established in a picosecond or so. The lifetime of the fluorescence is a few nanoseconds and thus laser dyes have nanosecond storage times. For generating ultra-short pulses it is necessary to employ a second absorber dye, which has a relaxation time, from the first excited singlet level to the ground state, shorter than the round trip time inside the laser cavity.

Figure 4 is a schematic diagram of a mode-locked dye laser. Regardless of the particular active medium employed, dye solution or doped solid, the standard laser resonator contains an absorber dye cell in contact with the 100% reflecting laser mirror. The saturable absorber has the essential property that it absorbs at the wavelength at which the laser medium fluoresces. With a dye laser the addition of an interferometer permits control of the lasing frequency and bandwidth and hence of the initial intensity fluctuation patterns. This turns out to be important for the generation of very short pulses. Such a mode-locked laser generates a train of pulses which are separated by the round trip time—the double transit time—of the laser resonant cavity. Figure 5 shows an oscillogram of a typical pulse train, recorded on a travelling-wave oscilloscope, with an instrument limited time resolution of about 300 psec. The pulses are very much shorter in duration. What happens inside the saturable absorber dye cell is indicated schematically in Fig. 6. Essentially, the initial photon noise interacts non-linearly with the saturable absorber dye cell. The weaker fluctuations are

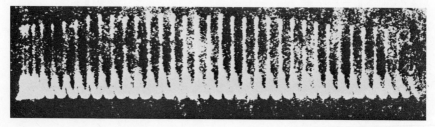

FIGURE 5. Mode-locked dye laser pulse train recorded on travelling wave oscilloscope.

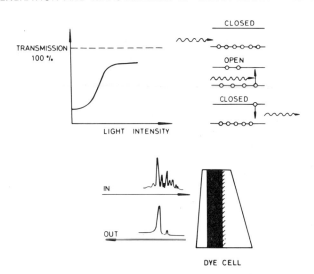

FIGURE 6. Schematic diagram showing effect of saturable absorber action on initial intensity fluctuation pattern.

absorbed, and the energy which is contained in these weaker fluctuations bleaches the dye solution. The result is that the absorption is a function of intensity, so that there is a nonlinear transmission function, with the transmission increasing with increasing light intensity until a saturation level is reached. The very large spikes are transmitted and the smaller ones are absorbed. Thus the largest fluctuation in the initial fluorescence intensity pattern of the laser medium is selected out and preferentially amplified. Figure 7 shows one of the useful properties of the dye laser compared with neodymium: glass or ruby systems. As a rough rule, a 2 psec pulse requires a 3Å bandwidth in the visible wavelength region. By employing a Fabry-Perot interferometer inside a dye laser cavity it is possible to select a particular portion of the bandwidth so that the laser not only produces very short pulses, but also generates them at a particular frequency. This then makes possible the study of resonant interactions. By using different dyes most of the visible spectrum is covered. As Professor Harris will tell us later, by employing nonlinear interactions, it should be possible to transpose a good picosecond pulse in the visible into the vacuum ultraviolet and even the soft X-ray region. Figure 8 is a photograph of some of the equipment showing a mode-locked laser oscillator and three amplifiers, of increasing diameter. With this system (which looks like a heart-and-lung machine because it is necessary to circulate the dye to maintain optical homogeneity) it is possible to generate 2 psec pulses throughout the visible region and amplify them to get peak powers greater than 3 GW.[3] That is about the present limit of dye laser systems.

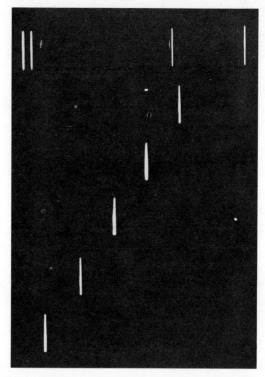

FIGURE 7. Spectra showing tuning of transform-limited picosecond pulse. Mercury calibration spectrum at top. Plate of 1 metre spectrograph was moved vertically as mode-locked dye laser was tuned from 603 nm to 625 nm. (From Ref. 3.)

FIGURE 8. Photograph of mode-locked dye laser oscillator and amplifier chain which produces ∼3 GW, 2 psec pulses.

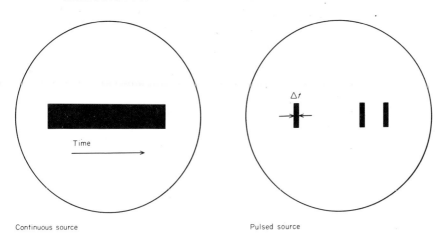

Continuous source — Pulsed source

FIGURE 9. Principle of electron-optical chronoscopy.

I have been talking confidently about the durations of these pulses. The question is: how do we measure pulse durations? You will have noticed that Dr. Pike's measurements stopped at somewhere around 200 psec. The reason for that is that for measurement using ordinary photomultipliers and fast photodiodes, the time resolution limit is ~ 100 psec. This is simply the limit of fast electronic circuits. To measure durations in the picosecond regime it is necessary to employ another approach. The solution has been around for something like 15 years. It is called electron-optical chronography, or more simply, chronoscopy. The principle of operation is based upon changing the time variable into a spatial variable, as illustrated in Figure 9. A slit image of the light source is focused on the photocathode of a streak image-tube. By applying a high-voltage ramp to the deflection plates the electron-optical image is moved across the tube phosphor at a velocity which approaches the velocity of light (3×10^{10} cm sec^{-1}). (You might worry about relativity, but with a bright enough torch I could move the beam on the surface of the moon with a speed even faster. So there is no problem there.)

If the slit is illuminated by a continuous light source, then a continuous track would be recorded. With ultra-short light pulses, slit images only are recorded as shown in Fig. 10. In order to measure duration one takes a ruler, and measures the track halfwidth $\Delta \ell$, and divides by the streak velocity to get the pulse duration Δt. There is nothing particularly novel about this technique. However, there is a fundamental problem in using streak cameras for picosecond measurements. The problem lies in the statistics of photoelectron emission from a photocathode. The energy distribution of the emitted photo-electrons has a finite spread. (See Fig 11.) As an example, at the neodymium laser wavelength of 1.06 μm, there is a spread of about

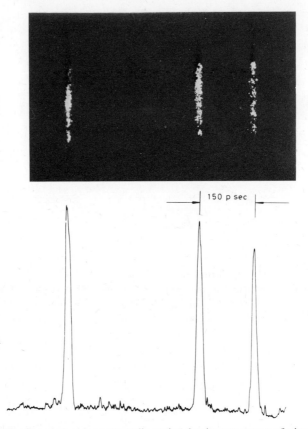

FIGURE 10. Streak record and corresponding microdensitometer trace of ultra-short pulses generated in optical delay line from a single pulse of a mode-locked neodymium: glass laser showing the first time resolution of <10 psec.

a third of an electron volt with an SI photocathode; at shorter wavelengths it is much greater. For that reason, up to four years ago the accuracy of measurement of ultra-short pulses was limited to about 60 psec, which was perfectly adequate before the advent of laser picosecond pulses. Basically what happens is that some of the photoelectrons are going faster than the others, and after a while the initial pulse spreads out. The solution of this problem was in fact remarkably simple. Figure 12 shows how time-dispersion, caused by the spread of electron velocities, is overcome[4] by generating a static high electric field close to the photocathode. This is very easily achieved in practice by the simple expedient[5] of placing a fine mesh electrode very close to the photocathode and applying a potential of a few kilovolts. The photoelectrons are accelerated up to a high velocity very quickly so that a spread of energy of a fraction of an electron volt does not matter.

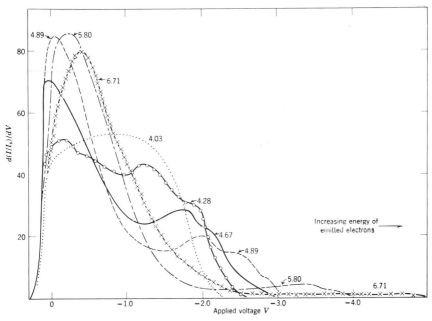

FIGURE 11. Distribution of photoelectrons emitted under illumination of light of different wavelengths.

Using this technique of increasing the extraction field—for instance, to get a resolution of about 1 psec, one needs something like 10 kV per centimetre[6]—it is quite easy to measure down to 1 psec in this manner. The typical experimental arrangement is also shown in Fig. 12. The pulse train, or a selected single pulse from the dye laser, or any other laser, is injected

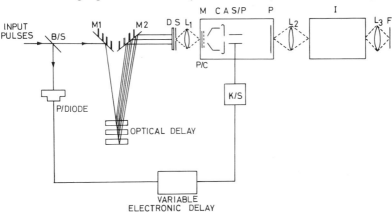

FIGURE 12. Chronoscopy showing optical delay line, streaking electronics, and details of camera system.

into an optical delay line. It is obviously important to know that measurements are accurate and it is very easy to calibrate the streak camera with a delay line. Consider one particular pulse. On reflection at the series of glass plates, the pulse is divided up into a series of pulses which are separated in space, and hence in time. This provides an automatic calibration of the streak records.

There is a problem, in the sense that if one tries to take too many photoelectrons in a very short time from a photocathode, it can become saturated; also space-charge effects occur. Both of these phenomena result in a distortion of the streaked image, with a consequent loss of time resolution. It is thus necessary to reduce the number of photoelectrons. This is most easily done by using an image intensifier tube. I should point out that this system then becomes a photon-counter because every photoelectron which passes through the extraction mesh can be recorded on the output screen of the intensifier system. The very rapid (nanosecond) kilovolt deflection potentials are obtained from electronic circuits based on avalanche diodes and Krytrons. The streak shown in Fig. 10 was obtained with our first picosecond camera.[4] One pulse was divided up into three pulses, delayed by calibrated known amounts. The pulse duration is accurately measured from the microdensitometer trace. This was one of the first shots showing a time-resolution of less than 10 psec.

Having developed the measurement techniques, it was then possible for us to confirm that we really knew what was going on inside a mode-locked

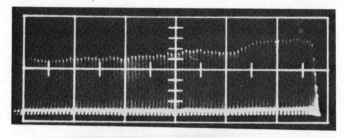

FIGURE 13. *Top*: Oscillogram of mode-locked pulse train from Rhodamine 6G dye laser tuned to operate at 605 nm. Time scale 50 nsec per major division.
Bottom: Streak camera record of "pulse" 18 nsec from start of train.

laser. To obtain the streak record of Fig. 13 the interferometer in the laser cavity was set[7] so that the bandwidth would correspond to the resolution of the camera. When a pulse near to the beginning of the train was recorded the photon noise structure was seen very clearly. In this way it was possible for us to study in detail the mechanism whereby a single fluctuation is selected out and all the energy is concentrated into it. Figure 14 shows examples of this.[8] Starting with the fluctuation pattern after 7 round trips (a), some of the larger spikes have been selected out (b), after 20 round trips. After 45 round trips, (d), one particular fluctuation has been selected and all the laser energy is concentrated into it. It is rather interesting that one can achieve the ideal situation, whereby the spectral bandwidth and the pulse-duration obey the uncertainty principle. This is called a transform-limited picosecond pulse, and for a dye laser the duration is of the order of a few picoseconds. We now know that in the dye laser there are two non-linear processes occurring.[7] These are saturable absorption which cuts away the leading edge of a pulse and saturable amplification which cuts off the trailing edge. Since two nonlinear processes are operating strongly together the mode-locked dye laser produces shorter pulses in a more reproducible way then mode-locked ruby and neodymium:glass lasers. Figure 15 shows a selection of streak photographs. These correspond to various types of laser and different types of mode-locking dye.[9]

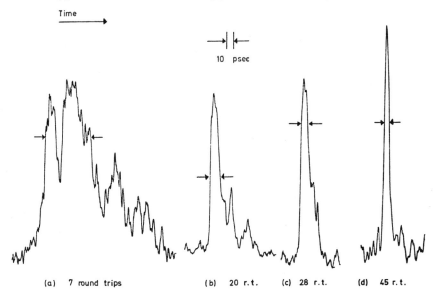

FIGURE 14. Sequence of microdensitometer trace of pulses after differing number of round trips. Half intensity points (linear density ordinate scale) of pulse envelopes are indicated by arrows.

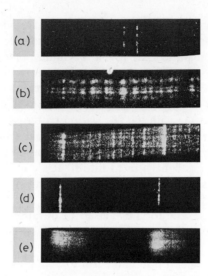

FIGURE 15. Streak records of pulses from different types of lasers (ruby and dyes) mode-locked by saturable absorbers. Note high quality and good signal-to-noise performance of streak camera.

What sort of experiments can one do with the combination of a mode-locked laser and very fast measurement techniques? The first obvious application is the measurement of fluorescence lifetimes. The particular example in Fig. 16 was obtained by exciting the mode-locking dye DODCI with a mode-locked dye laser and time-resolving the resultant fluorescence.[7] By microdensitometry a direct measurement of the fluorescence lifetime is obtained. The lifetime of this dye is 250 psec. I should point out that this value was obtained under mode-locking conditions. It is not quite the value that would be obtained if measured outside the mode-locking cavity. Stimulated emission and photoisomer generation influence the lifetimes inside the laser cavity.[10] Time-resolved spectroscopy can also be obtained

FIGURE 16. Example of fluorescence lifetime measurement (DODCI) by electron-optical chronoscopy.

with the streak camera system, with dispersion along the slit. In this way the full information content obtainable from a photographic plate would be employed.

One of the important roles of physics is to be the tool-making science. This task is really never-ending, because as soon as a new measurement technique is developed, somebody wants to measure the property a bit more accurately. A particular example of this is the use of mode-locked neodymium: glass lasers for laser compression, with the aim of controlled fusion. When a laser pulse becomes very intense, nonlinear effects occur inside the laser glass rod or disk, due to the fact that the optical frequency electric field is itself affecting the refractive index. This is known as self-phase modulation, which produces a modulation of both the frequency and intensity envelopes of the laser pulse. Thus the pulse begins to develop sub-structure. It turned out that the first picosecond streak cameras did not have adequate time resolution to study this phenomenon in detail. In the last year or so at Imperial College we have pushed the technique of chronoscopy to what, I think, is approaching the limit of electron-optical technology. There are other reasons for doing this, particularly in photochemistry and photobiology as Fig. 17 shows. Curiously enough, 1 or 2 psec seems to be a fairly significant time. A picosecond is about the time period during which a molecule in a condensed phase collides with its environment. In photochemistry most intramolecular and intermolecular vibration relaxation times, have rates of $\sim 10^{12}$ per second.

So it seemed worthwhile to us to try to push the measurement technique into the sub-picosecond region so that it would be possible not to be limited by camera time-resolution when investigating the excited states of molecules. The other motive was that if XUV and X-ray lasers are going to work, and if we get ultra-short pulses from these shorter wavelength lasers, then the time-scale changes from picoseconds to femtoseconds. The reason is that the lifetimes of these laser transitions are femtoseconds. Of course with these very short wavelengths there are sufficient optical cycles to define shorter pulses. So we felt we would push the measurement technique as far as we could, and Fig. 18 shows the latest version of the image tube. The improvement essentially consisted of redesigning the photocathode end to produce a higher electric field there of 20 kV cm^{-1}. The overall electron-optics were also improved to increase the spatial resolution to 10 ℓ/psec mm^{-1} in streak mode. This produces a doubling of the information content and, incidentally increases the overall recording speed by a factor of four, so that correspondingly weaker light sources may be studied. This device is now capable of measuring down to a tenth of a picosecond and more importantly, of having a time resolution of less than 2 psec over all the spectrum from 120 nm to 1200 nm. In this field, one is either ahead

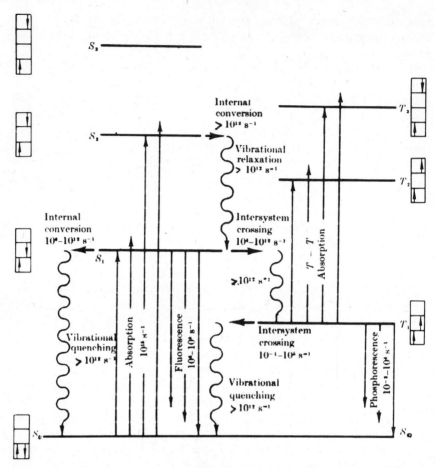

FIGURE 17. Molecular relaxation rates.

FIGURE 18. Photograph of Photochron II streak camera tube.

GENERATION AND MEASUREMENT OF ULTRA-SHORT PULSES 105

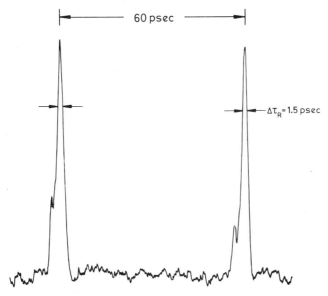

FIGURE 19. Microdensitometer trace of a pair of streak images of a pulse generated by a Rhodamine 6G dye laser, mode-locked using DQOCI and showing a recorded pulse duration of 1.5 psec.

in producing shorter duration light sources or in the measuring technique. Just now the difficulty is to find a light source which is of short enough duration to test the limits of the Photochron II tube. Figure 19 shows some results obtained recently by Dr. Sibbett at Imperial College. The pulse images are separated by 60 psec and the recorded width is 1.5 psec. Occasionally in the initial laser intensity fluctuations two spikes occur close together which are of about equal intensity. If one is very lucky, and takes enough shots, one might get a streak record which would show perhaps a double-peaked pulse separated by one or two picoseconds, which would provide a measurement of the camera instrumental function. After many trials such a record was obtained (Fig. 20). This represents a laser firing with the right initial fluctuation pattern to produce the desired test pulse shape. The recorded width of each peak is one picosecond, which tells us that the camera instrumental width is half a picosecond. It seems now that there is no problem in measuring the ultra-short pulses which are likely to be generated from present laser sources.

What is the most likely way of achieving the ultimate limit of pulse duration allowed by the laser bandwidth? I think it is to use the continuous-working dye laser. I also think that this dye laser is the one which will be of most use in generating ultra-short pulses for investigations in other branches of science. With a continuous pulse train it is easier to apply all the other well

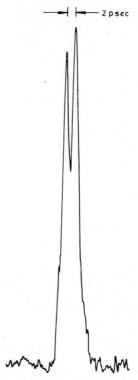

FIGURE 20. Microdensitometer trace of dye laser pulse with substructure separated by 2 psec and showing a total camera resolution limit of <1 psec.

known techniques of spectroscopy such as phase sensitive detection and optical beats, apart from the considerable convenience in alignment and calibration. Figure 21 shows the arrangement employed in our laboratory for a continuous working mode-locked dye laser.

A powerful Argon ion laser of up to 20 W power is focused to a spot size of 10 μm in the dye cell. The dye laser beam is similarly focused into the saturable absorber cell. The prism tunes the output frequency and by this technique we get a continuous train of pulses of about 2 psec duration. Figure 22 shows an oscillogram of the pulse train recorded on a pulse-sampling oscilloscope. The time-resolution is inadequate and to determine true pulse durations it is necessary to employ either nonlinear interferometer techniques or the streak camera. Employing both of these methods we have confirmed that pulse durations of 2 picoseconds are reliably obtained.

I should like to finish by briefly discussing the X-ray region. In the early days of lasers, it was quite common to amuse visitors by focusing the beam of a pulsed ruby or neodymium laser onto a point in the air. The

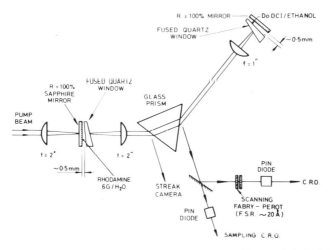

FIGURE 21. Experimental arrangement of a continuous working mode-locked dye laser.

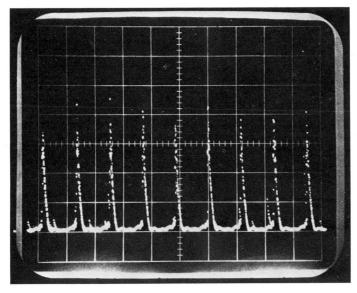

FIGURE 22. Mode-locked train of 2 psec pulses from a tunable frequency c/w dye laser.

result would be a loud bang and a very bright spark. As we now know that was probably one of the more interesting phenomena which were happening in laser laboratories, because these laser sparks are very hot plasmas. If a solid target such as iron or lead is employed, this produces a very copious point source of X-rays and the energy in ultra-short pulses, of duration of a few nanoseconds or less, from high powered lasers in the infrared can be converted with about 30% efficiency into X-rays.[11] These intense point sources have immediate applications in medical photography and possibly for X-ray therapy.

It is also fairly clear that the most likely path to the production of an X-ray laser, is the use of high-power ultra-short laser pulses for pumping. In collaboration with Dr. Key at Queen's University, Belfast[12] we have been developing diagnostic techniques for studying very hot laser plasmas on a picosecond time scale. Figure 23 shows the experimental arrangement. A pulse from a high-power Nd:glass laser was focused onto a copper target. The plasma X-ray source had dimension 50 µm × 100 µm and emitted strong resonance line radiation particularly of CuXX, in the 1 keV region with an underlying recombination continuum. The plasma was imaged on to the photocathode with thin aluminium filters to select out the X-radiation. With pinhole optics (25 µm pinhole) a spatial resolution of 40 µm was achieved. With the experimental arrangement shown in Fig. 24, streak records were readily obtained and confirmed that high spatial and temporal resolution can be obtained with these intense X-ray point sources (Fig. 25).

Already in photochemistry and photobiology, in laser compression and in studies aimed at XUV and X-ray lasers, picosecond pulses are playing a dominant role. A very interesting and unexpected application is for satellite radar, whereby it is possible to measure continental drift to a few centimetres simply by firing a pulse up from one continent, reflecting it from a satellite and using a streak camera on a second continent. More recently, it has been suggested that this technique be used for earthquake detection or prediction by monitoring the movement of the earth surface, which changes in altitude before an earthquake.

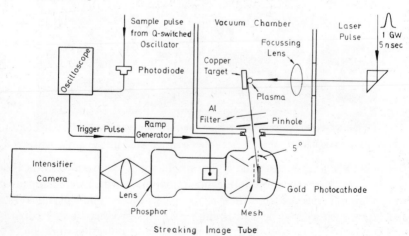

FIGURE 23. X-ray streak camera experimental arrangement.

FIGURE 24. Photograph of X-ray streak image-tube and associated apparatus.

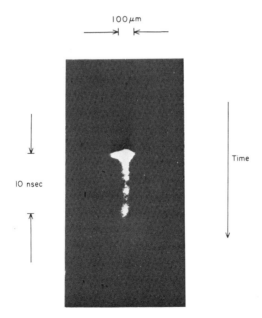

FIGURE 25. X-ray streak record showing spatial resolution of 40 μm and source limited time resolution >150 psec.

REFERENCES

1. Nuckolls, J., Wood, L., Thiessen, A. and Zimmerman, G. (1972), *Nature*, **239**, 139–142.
2. For a good review of dye lasers see "Dye Lasers", Schäfer, F. P. ed., Springer-Verlag Berlin (1973).
3. Arthurs, E. G., Bradley, D. J. and Roddie, A. G. (1971), *Appl. Phys. Lett.*, **19**, 480–481.
4. Bradley, D. J., Liddy, B. and Sleat, W. E. (1971), *Optics Communs*, **2**, 391, 395.
5. Bradley, D. J. US Patent 3761614.
6. Bradley, D. J. and New, G. H. C. (1974), *Proc. IEEE*, **62**, 313–345 and references therein.
7. Bradley, D. J. (1974), *Opto-electronics*, **6**, 25–42 and references therein.
8. Adrain, R. S. (1974), PhD thesis, Queen's University, Belfast.
9. Arthurs, E. G., Bradley, D. J. and Glynn, T. J. (1974), *Optics Communs*, **12**, 136–139.
10. Arthurs, E. G., Bradley, D. J., Puntambekar, P. N. and Ruddock, I. S. (1974), *Optics Communs*, **10**, 360–365.
11. Key, M. H., Eidmann, K., Dorn, C. and Sigel, R. (1974), *Appl. Phys. Lett.*, **25**, 335.
12. Bird, P. R., Bradley, D. J., Roddie, A. G., Sibbett, W., Key, M. H., Lamb, M. J. and Lewis, C. L. S. (1975), "Proceedings of the 11th Congress on High Speed Photography", Chapman and Hall, 118.

Chapter 6

Ultra-short Pulse Interaction Studies

by W. Kaiser, Technische Universität München, 8 München 2,
Arcisstrasse 21, West Germany

With present experimental techniques it is possible to generate reproducibly light pulses of a time duration of several 10^{-12} sec. In laser systems working with Nd-doped glass as an active material, these ultra-short pulses are immediately produced at very high light intensity of the order of 10^8 W/cm^2. On account of this high intensity, detection and analysis of ultra-short pulses is readily possible. In addition, a series of interaction processes between electromagnetic radiation and matter has been observed and investigated in recent years. In fact, it was found to be difficult to transmit these very high power pulses over longer distances (especially in condensed matter) because of various strong interaction processes. The following examples should elucidate this point:

(1) intense monochromatic light pulses readily generate new frequencies; complicated spectra have been observed after an interaction length of less than 1 cm;

(2) strong deviations from the original intensity distribution over the cross section of the beam have been observed on account of self focusing action of the electromagnetic pulse;

(3) dielectric breakdown with rapid generation of free electrons has been found in solids, liquids and compressed gases, when the light pulse exceeds a certain field strength.

A new branch of physics, nonlinear optics, has developed which deals with these various light-matter interaction processes. Many hundred papers were devoted to this active field of modern science during the past decade. Several unexpected interaction processes were found with interesting applications in other fields.

Nonlinear optics starts with the fact that the dielectric constant ε is not constant any longer for very high values of the field strength of the electromagnetic wave, i.e. ε becomes field dependent, $\varepsilon = \varepsilon(E)$. As a result, the total polarization of the medium is a complicated function of the electromagnetic field. The following terms are of special interest here.

$$P = \chi' E + \chi'' E E + \chi''' E E E$$

The first term represents the proportionality of polarization and field which everybody is familiar with since their undergraduate days. The second and third terms have important consequences; e.g. when the electromagnetic fields are taken to be of equal frequency ω then a polarization at 2ω and 3ω occurs which, under favourable conditions, can radiate electromagnetic waves at second and third harmonic frequencies. In other papers of this conference nonlinear optical processes are discussed which result from these higher terms of P. For instance, the infrared spin-flip Raman laser of S. D. Smith is based on a third-order process as is the third harmonic generation of ultraviolet radiation presented by S. E. Harris.

In this paper I wish to discuss two nonlinear interaction processes; both require ultra-short light pulses. The first subject deals with parametric interactions of second order where $P_i = \chi'' E_S E_L$; i.e. a polarization at an idler frequency is generated by the interaction of a signal and a laser field. The second topic is concerned with the generation of intense coherent vibrational excitations. A Stokes shifted field E_S together with the laser field E_L drive a coherent material excitation of amplitude Q at the difference frequency i.e. $Q \propto \chi''' E_S E_L$. In this Raman type process a radiating polarization at $P_S \propto Q E_S$ is produced which is the source of the electromagnetic Stokes wave.

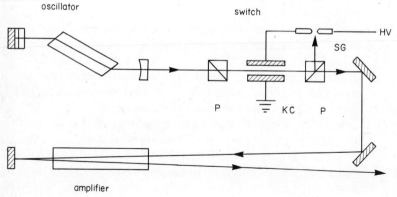

FIGURE 1. Schematic of a system to generate single ultra-short light pulses.

Pulse duration	$t_p = 6$ psec
Contrast radio	3 ± 0.1
Frequency width	$\Delta v = 2.9$ cm^{-1}
$t_p \times \Delta v =$	0.6
Peak to background ratio	$\sim 10^4$
Pulse intensity	5×10^8 W/cm^2
Pulse energy after amplification	4×10^{-3} J

FIGURE 2. Relevant parameters of our single ultra-short light pulses.

After these introductory remarks let me turn to our experimental systems and our various results. In Fig. 1 a schematic of our generator for ultra-short light pulses is depicted. The mode-locked laser oscillator generates a train of ultra-short pulses. It has been well established in recent years that pulses within the pulse train are not identical; they vary in pulse width, in frequency spectrum and in pulse shape. As pointed out in the paper by D. G. Bradley the pulse shape deteriorates drastically at the end of the pulse train. In order to have reproducible pulses of approximately Gaussian time dependence it is necessary to cut one pulse out of the early part of the pulse train with an electrooptic switch. A following optical amplifier increases the intensity of the single pulse to the desired value. In Fig. 2 the essential parameters of our ultra-short pulses are summarized. We work with a pulse duration of $t_p = 6$ psec, we have a measured frequency width of $\Delta v = 2.9$ cm^{-1} which gives a band width limited pulse as seen from the product $t_p \times \Delta v = 0.6$. Note, that for a Gaussian pulse a product of 0.45 is calculated. According to our experience it is essential to work with well defined single picosecond pulses. Satellite pulses and pulses with complicated subpicosecond structure have given poor and even erroneous results. Our pulse intensities are of the order of several 10^8 W/cm^2 and the pulse energy is several 10^{-3} J or several 10^{16} photons per pulse.

Let me turn now to the optical parametric process as briefly outlined above. In this nonlinear optical interaction the incident pump pulse at frequency ω_p generates two pulses at the signal frequency ω_s and the idler frequency ω_i. According to Fig. 3, these three electromagnetic waves have to follow the energy conservation law $\omega_p = \omega_s + \omega_i$ and they have to fulfil the phase

Optical parametric process

$\omega_p = \omega_s + \omega_i$

$k_p = k_s + k_i$

FIGURE 3. Energy conservation and phase matching conditions in the optical parametric process.

Parametric generator

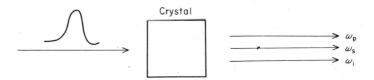

Tuning: crystal orientation, temperature

FIGURE 4. Schematic of an optical parametric generator.

matching condition $\mathbf{k}_p = \mathbf{k}_s + \mathbf{k}_i$. The question now arises what determines the frequency ω_s or ω_i. The magnitude of the wave vector $\mathbf{k} = \omega n/c$ is determined by the frequency and the index of refraction of the medium n; the latter is a function of frequency (colour dispersion) and direction in an uniaxial crystal (index ellipsoid). As a result, well defined directions between the electromagnetic beam and the crystal axis can be calculated where phase matching is perfect and strong parametric interaction is observed. At other angles a phase mismatch gives in most cases negligible parametric signals. The parametric generator is schematically depicted in Fig. 4. A powerful light pulse travels through a crystal which, when properly oriented, gives, apart from the transmitted input wave, two new desired pulses at frequency ω_s and ω_i. Tuning can be accomplished by crystal orientation and in some cases (not for our infrared pulses) by temperature variation. In the case of a parametric oscillator two mirrors on both sides of the crystal strongly increase the interaction between the pump light and the crystal. For ultrashort light pulses of a geometrical length of less than one millimetre the optical feed back mechanism of two oscillator mirrors does not work. In Fig. 5 a simplified expression of the expected signal and idler intensity is presented. It is readily seen that the signal depends strongly on the magnitude of $\gamma z I_p^{\frac{1}{2}}$. For the product $\gamma z I_p^{\frac{1}{2}} < 1$, the sinh function is readily expanded with the signal intensity being proportional to $\gamma z I_p^{\frac{1}{2}}$. In this domain of small gain, extensive investigations have been made in recent years. The small signal

$$\text{Signal: } I_s \sim \sinh^2(\gamma z \sqrt{I_p})$$

$$\text{Idler : } I_i \sim \cosh^2(\gamma z \sqrt{I_p})$$

$$\gamma = \left[\frac{32\pi^3}{c^3} \frac{\omega_i \omega_s}{n_i n_s n_p}\right]^{1/2} \chi_{\text{eff}}$$

FIGURE 5. Signal and idler intensity for parametric fluorescence.

output makes the parametric fluorescence a very inefficient and impractical light source. Quite different is the situation when the pump intensity is high enough, that $\gamma z I_p^{\frac{1}{2}} > 1$. In this case the signal grows exponentially with $\gamma z I_p^{\frac{1}{2}}$ giving rise to substantial conversion efficiency as seen below. Before going to the next figure a short remark should be made concerning the gain coefficient γ of Fig. 5. The most important parameter of γ is the nonlinear susceptibility χ_{eff} which has been measured for a variety of crystals in recent years. Knowing this number, and in this way γ, it is possible to predict the total gain beforehand for a certain material.

In Fig. 6, the tuning curve of a $LiNbO_3$ crystal is presented. The solid curve is calculated from the indices of refraction of the material; our experimental points (open circles) are in perfect agreement with this curve. In this material we are able to generate with an input pulse at 1.06 µm, signal and idler pulses between 1.4 µm and 3 µm.

Most interesting is the result on the conversion efficiency of our ultrashort light pulses. In Fig. 7 the signal power at $v_s = 6500 \text{ cm}^{-1}$ is plotted as a function of input peak intensity. One readily sees the exponentially rising signal intensity with a slope which is in good agreement with the material parameter χ_{eff} of the nonlinear crystal. The signal intensity appears to

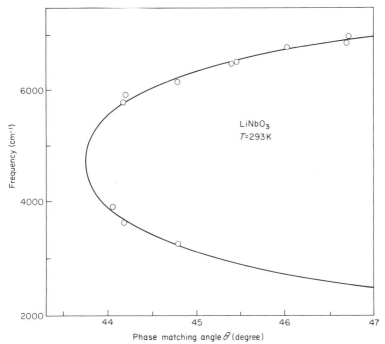

FIGURE 6. Tuning curve of a $LiNbO_3$ crystal.

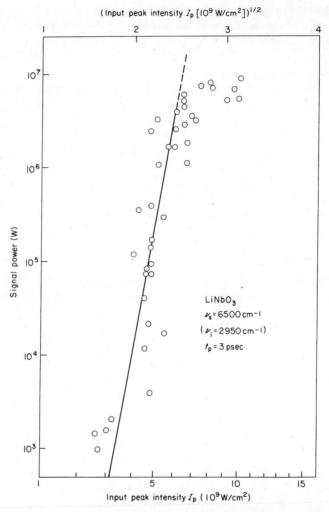

FIGURE 7. Signal power versus peak intensity of input pulse for parametric fluorescence.

saturate at an intensity conversion of several per cent making this simple device a very practicable source of tunable intense picosecond pulses. Two points should be noted. First, the parametric process discussed here is not restricted to the infrared part of the spectrum. Very recently, S. E. Harris has operated a similar system in the visible. Second, further improvement of the signal output is possible using two crystals in sequence with the crystals separated by approximately 50 cm. In this way the angular width and the frequency spread of the signal pulse is reduced.

The ultra-short tunable pulses should be useful for a variety of physical applications in the infrared. In particular molecular vibrations in condensed matter are readily excited with these short pulses allowing the study of energy relaxation times in a large number of molecular transitions, where the dipole moments are sufficient for direct excitation. We have started a programme in this direction and our preliminary data look most promising. In fact, we have seen the decay of excess vibrational energy of the CH-stretching mode in CH_3J after direct excitation with an ultra-short light pulse at 3 μm.

The next part of my paper is concerned with the stimulated Raman process, a third-order interaction in the electric field as discussed above. Stimulated Raman scattering can be visualized as follows (see Fig. 8). A powerful incident electromagnetic wave is scattered on a material excitation of the medium producing strong Stokes-shifted radiation. These two waves generate, via a nonlinear interaction term, more material excitation which in turn gives rise to enhanced Stokes scattering. As a result, the Stokes radiation grows

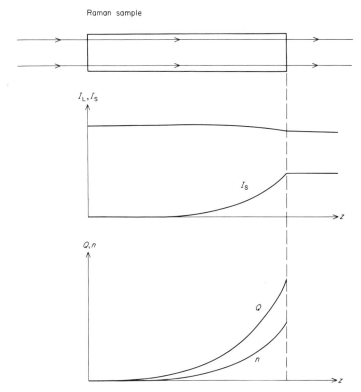

FIGURE 8. Schematic of the generation of the Stokes intensity and the material excitations Q and n in the stimulated Raman process.

exponentially with distance, with substantial conversion of laser into Stokes light. In most previous investigations of this subject emphasis has been given to the analysis of the frequency shifted electromagnetic radiation. During the past year we focused our attention on the material excitation which occurs in the interaction process. Since the incoming laser light and the generated Stokes emission are coherent electromagnetic waves the material excitation will be coherent with an amplitude Q rising rapidly as a function of distance in the medium (see lower part of Fig. 8). There is a second type of material excitation. In the Raman process one molecule is promoted from the ground state to the first vibrational state for each Stokes photon emitted. Since in stimulated light scattering large photon fluxes are involved, the material excitation is substantial, e.g. with the ultra-short light pulses discussed above (Fig. 2) it is possible to produce an occupation of $n \simeq 10^{-3}$. One out of a thousand molecules is promoted to the first vibrational state within the duration of the short pulse interaction. This occupation density is for molecular modes of high vibrational energy, far above the thermal equilibrium value.

I wish to stress the difference between the coherent amplitude Q and the occupation number n. Both grow rapidly during the interaction process between the electromagnetic pulse and the medium. After the pulse has travelled through the medium the two excitations have different decay times on account of different physical processes which effect the two material excitations. Coherent excitation means that there is a definite phase relation between vibrating molecules at various positions within the medium. Resonant transfer of vibrational energy from one molecule to another leads frequently to a loss of phase during this transfer process. Loss of phase and loss of coherence means a decay of Q. On the other hand, the transfer of vibrational energy does not change the occupation number n. The coherent amplitude might disappear in the excited volume but the vibrational energy still persists. We have studied these two relaxation processes and the rest of this paper will be concerned with the techniques and with results of such investigations.

Figure 9 shows the differential equations which govern the stimulated Raman process. This figure is presented in order to explain once more the two material excitations and their respective relaxation times. In the first line of Fig. 9 the wave equation of the total electromagnetic field (E_L and E_S) is given. On the right hand side the nonlinear polarization represents a driving term for the interaction. The polarization is directly related to the change of polarizability with normal coordinate $\partial \alpha / \partial Q$ which is characteristic for Raman type interactions. The second and third line represent material equations for the coherent amplitude Q and the occupation number n, respectively. Note the different relaxation times for Q and n. The time constant τ is called the dephasing time since it is connected with the loss of

$$\frac{\partial^2 E}{\partial z^2} - \frac{1}{c^2}\frac{\partial^2(\varepsilon E)}{\partial t^2} = \frac{\pi}{c^2}\frac{\partial^2 P}{\partial t^2}$$

$$P = N\left(\frac{\partial \alpha}{\partial Q}\right) Q E$$

$$\frac{\partial^2 Q}{\partial t^2} + \frac{1}{\tau}\frac{\partial Q}{\partial t} + \omega_0^2 Q = \frac{1}{2m}\left(\frac{\partial \alpha}{\partial Q}\right) E^2 (1 - 2n)$$

$$\frac{\partial n}{\partial t} + \frac{1}{\tau'} n = \frac{1}{2\hbar\omega_0}\left(\frac{\partial \alpha}{\partial Q}\right) E^2 \frac{\partial Q}{\partial t}$$

FIGURE 9. Differential equations governing the stimulated Raman process.

phase of the vibrating molecules. The relaxation time τ' is a measure of the decay of excitation of the first vibrational state; with this time constant the vibrational energy of the excited mode decays to lower excited states or in some cases directly to the thermal bath.

How do we study these two material excitations and how do we measure the two relaxation times? We certainly require different experimental systems to determine the generated coherent excitation and the occupation density. In Fig. 10 two experimental techniques are depicted schematically. In both cases the material is coherently excited with a first powerful laser pulse of frequency ω_{L1}. The relaxation times are measured with a probing technique where a second ultra-short light pulse of frequency ω_{L2} passes with a well

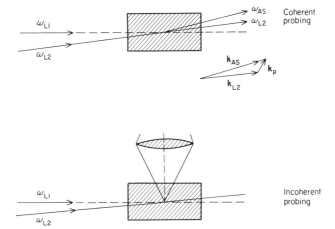

FIGURE 10. Schematic of the experimental probing techniques to determine the dephasing and the energy relaxation time.

defined delay time through the excited volume and probes the degree of excitation. Two situations should be distinguished. In the case of coherent probing the coherent excitation is under study, i.e. the phase of the excitation is significant. As a result, the light scattered off the material excitation has to be measured under well defined phase matching conditions. The insert in Fig. 10 shows a phase matching triangle between the wave vectors of the material excitation k_p, the probe pulse k_{L2} and the scattered wave k_{AS}. Calculations show that the coherent scattering signal occurs close to the forward direction with angles of less than a few degrees. Substantially different is the situation for the investigation of the degree of excitation of the first excited state (incoherent probing). It is well known that spontaneous anti-Stokes scattering is directly proportional to the occupation of the first vibrational state. It is convenient to measure this type of scattering under an angle of 90°. The incoherent probe scattering gives very small signals on account of the small spontaneous Raman scattering cross sections. Fortunately, the substantial excitation possible with powerful pumping pulses gives enough signal for a practical application of this technique.

Figure 11 serves to illustrate the probing technique. The dotted curve represents the incident pulse I_L as a function of time. This pulse generates the coherent excitation Q which builds up in a transient manner reaching its maximum value on the decaying side of the pump pulse. It is important to note that the coherent material excitation decays slowly with its own characteristic time constant τ while the incident laser pulse of pulse duration t_p

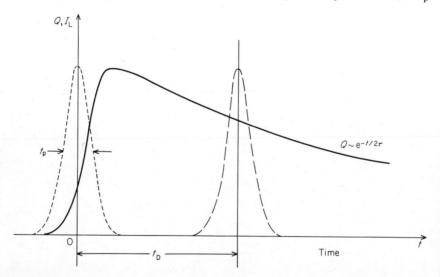

FIGURE 11. Schematic of the probing technique. Probe pulse arrives after delay time t_D at the excited volume.

decays rapidly (approximately with a Gaussian function). A second pulse (broken line) arrives at the excited volume with a time delay t_D. The coherent scattering signal $S(t_D)$ is a convolution of the probing pulse and the coherent excitation after time t_D. It can be shown that the signal $S(t_D)$ decreases as a function of delay time t_D with the time constant τ.

In Fig. 12 experimental data of the dephasing time of the symmetric normal vibration of CCl_4 is presented. The coherent scattering signal shows a clear exponential decay with a time constant of 4 psec. Similar experiments of different normal modes of various molecules gave dephasing times varying over approximately two orders of magnitude. In Fig. 13 our experimental relaxation times are compared with time constants which were deduced from the spontaneous Raman line width of the same normal mode. As seen from the figure, there is excellent agreement between the two time values for each molecule. As a conclusion from our findings we can state that the Raman line width is determined by the dephasing time of the normal modes under discussion; no information concerning the energy relaxation is obtained from spontaneous Raman data.

After discussion of the dephasing time I would like to move on to the energy relaxation time τ' introduced above. An example of our experimental

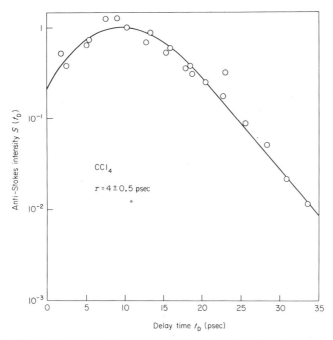

FIGURE 12. Coherently scattered signals versus delay time of probe pulse. Example of the determination of the dephasing time. The totally symmetric vibration of CCl_4 is measured.

Dephasing time

	ν_0 (cm^{-1})	τ (10^{-12} sec)	
		from line width	measured
CH$_3$CCl$_3$	2939	1.1 ± 0.1	1.3 ± 0.7
CCl$_4$	459	4.0 ± 0.5	4.0 ± 0.5
N$_2$(77K)	2326	79 ± 8	75 ± 8

FIGURE 13. Comparison of experimentally determined dephasing times with time values calculated from the spontaneous Raman linewidth.

results is presented in Fig. 14. A CH-stretching mode of ethanol around 2900 cm^{-1} is first excited by stimulated Raman scattering and the (incoherent) spontaneous anti-Stokes signal is plotted as a function of the delay time of the probing pulse with respect to the pumping pulse. The solid curve of Fig. 14

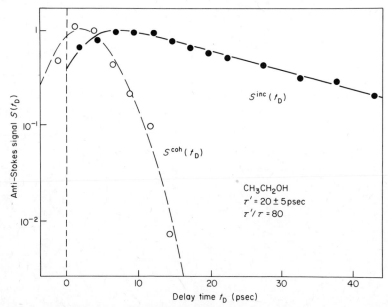

FIGURE 14. Incoherently scattered signals (full points) versus delay time of probe pulse. Example of the determination of the energy relaxation time. The CH-stretching vibration of CH$_3$CH$_2$OH is investigated.

shows a slow decay of the excess population with a time constant of $\tau' = 20$ psec. This number has to be contrasted with the very rapid dephasing of the same normal mode of the same molecule. The loss in phase is so fast in ethanol that we are unable to determine the value with our pulses of several psec duration. The number for τ (estimated from the spontaneous Raman line width) indicates that the loss of phase occurs approximately 80 times faster than the loss of excess population.

The question now arises of which physical processes determine the energy relaxation times discussed in the previous chapter. In particular, is there efficient vibrational–vibrational energy transfer between different molecules in the liquid, or in other words, do intermolecular or intramolecular processes determine the decay of vibrational energy? In order to shine some light on these important questions concerning the dynamics of the liquid state we have performed a series of experiments. Two relevant results will be presented in the following two figures. In Fig. 15 the energy relaxation time of the CH stretching vibration of CH_3CCl_3 at $\nu_H = 2939 \text{ cm}^{-1}$ is investigated: CH_3CCl_3 is diluted with CCl_4. In pure CH_3CCl_3 the energy relaxation is found to be fast with a time constant of $\tau' = 5.2$ psec. With increasing addition of 20, 40 and 60 mole per cent of CCl_4 the energy relaxation rises by approximately a factor of six to a value of $\tau' = 29$ psec. This strong rise of

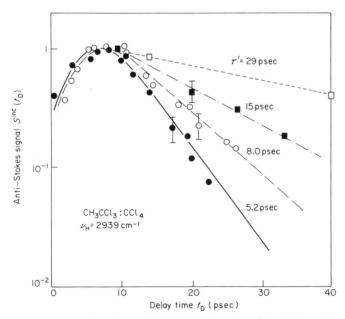

FIGURE 15. Concentration dependence of the energy relaxation of the CH-stretching mode of CH_3CCl_3 in mixtures with 0, 20, 40 and 60 mole percent of CCl_4.

τ′ differs drastically from a small change (10%) of the Raman linewidth of the same liquid mixtures. This result stresses once more the previous statement that measurements of the Raman line width give no information on the energy relaxation times.

When the values of τ′ are plotted versus concentration a quadratic dependence is found which suggests an interaction mechanism with neighbouring molecules. More direct evidence for the transfer of vibrational energy between molecules in the liquid state is obtained from experimental data depicted in Fig. 16. In the mixture of CH_3CCl_3 and CD_3OD (60:40) the CH vibration at approximately 2900 cm^{-1} is excited first via stimulated Raman scattering of the pump pulse. The spontaneous anti-Stokes signal (open circles) with a frequency shift of 2900 cm is a measure of the growth and decay of this vibration (broken line). Tuning the spectrometer to a frequency shift of 2200 cm^{-1} we are able to find an anti-Stokes signal (solid points) which has a completely different time dependence. At 2200 cm^{-1} there is a v_D vibration of CD_3OD which, according to our experimental results, is populated during the decay of the 2900 cm^{-1} vibration of CH_3CCl_3. In fact, the solid curve drawn through the solid points was calculated on the

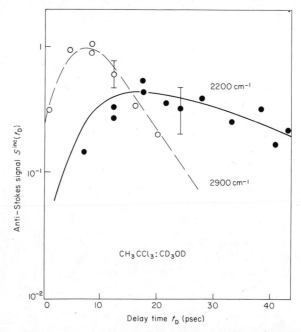

FIGURE 16. Energy transfer between the CH_3CCl_3 and the CD_3OD molecule. The v_H-vibration of CH_3CCl_3 at 2900 cm^{-1} is first excited by the pump pulse. The subsequent rise and decay of the occupation of the v_D-mode of CD_3OD at 2200 cm^{-1} is measured (full points).

basis that vibrational energy from the CH_3CCl_3 molecules is transferred to CD_3OD with an efficiency of 60%. I believe that this experiment gives convincing evidence of efficient energy transfer in liquid systems. It is important to emphasize that in the mixed system discussed here a kind of resonance process occurs since the v_D vibration of CD_3OD plus a vCl vibration of CH_3CCl_3 add up quite accurately to the primary energy of the v_H mode.

In conclusion I wish to say that the high field strengths available with ultra-short pulses allow us to produce strong interactions between the electromagnetic wave and the material. The short time duration of the pulses is ideally suited for studies of processes with time constants of the order of 10^{-12} sec. Molecular vibrations in liquids or optical phonons in solids have relaxation times in this time domain. The experiments presented here are examples of the successful application of ultra-short pulses for the study of very fast molecular processes.

ACKNOWLEDGEMENTS

The author gladly acknowledges the important contributions by Dr. A. Laubereau.

General References

The following publications give further information on the subjects discussed in this article.

Laubereau, A., Greiter, L. and Kaiser, W. (1974), *Appl. Phys. Lett.,* **25,** 87.
Von der Linde, D., Laubereau, A. and Kaiser, W. (1971), *Phys. Rev. Lett.,* **26,** 954.
Laubereau, A., von der Linde, D. and Kaiser, W. (1972), *Phys. Rev. Lett.,* **28,** 1162.
Laubereau, A., Kirschner, L. and Kaiser, W. (1973), *Optics Communs,* **9,** 182.
Laubereau, A., Kehl, G. and Kaiser, W. (1974), *Optics Communs,* **11,** 74.

Chapter 7

Coherent Optical Spectroscopy

by R. G. Brewer, IBM Research Laboratory, San Jose, California, U.S.A.

INTRODUCTION

Molecules exposed to a beam of laser light can exhibit interesting and unusual radiative characteristics. The coherent nature of the optical field forces the molecules to respond in unison during excitation and furthermore, to emit in unison, at some later time, a beam of coherent light. The coherence of the laser field is transferred, therefore, to the molecular sample in a type of preparation that determines its transient emission behaviour in the future.

Until recently, this class of optical coherence phenomena has not yielded readily to observation. The method of "Stark-pulse switching"[1] developed in our laboratory in 1971 provides, however, a simple way of exploring molecular coherence effects of this kind. Instead of using pulsed laser light, as in previous experiments,[2,3] the laser amplitude and frequency remain fixed and the molecular transition frequency is shifted in or out of resonance with that of the laser by means of a pulsed dc Stark field. (See Fig. 1) A coherent transient absorption or emission signal then follows. The technique is simple and yet versatile because it involves the use of conventional electronics for generating prescribed electric (Stark) pulse sequences. By varying the pulse sequence, any of a number of coherent transient experiments can readily be performed.

This article discusses three current applications of coherent optical transient phenomena:

    (1) the identification and study of specific molecular collision mechanisms which until now have remained hidden within the optical lineshape;

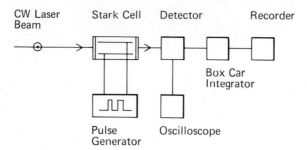

FIGURE 1. Method for observing coherent optical transients following a sequence of Stark pulses. (from Ref. 1)

(2) optical pulse Fourier transform spectroscopy, the optical analog of pulsed NMR, where exceptionally high spectral resolution, in the range 10^8–10^9 or higher, is possible without the complication of Doppler broadening;
(3) the generation of coherent optical pulse trains that can be electronically manipulated and where the time scale can be compressed continuously by varying the Stark field amplitude.

THE STARK-SWITCHING TECHNIQUE

Figure 1 is a schematic of the Stark-switching apparatus[1] that has now allowed the observation of optical nutation,[1] photon echoes involving two[1] or multiple pulses (stimulated and Carr-Purcell echoes[4]), optical free induction decay[5] (FID), coherent Raman beats,[6] optical adiabatic fast passage,[7,8] the optical analogue of spin locking[9] and FID interference pulses.[10] With the exception of Raman beats, all of these effects are the optical analogues of spin transients.[11-15]

A molecular gas sample that is Stark tunable is irradiated by a continuous wave CO_2 laser beam. Electronic pulses are applied repetitively to the sample thereby switching the molecular level structure in or out of resonance with a fixed laser frequency, and a particular coherent transient effect can be selected by simply varying the pulse sequence.

In Fig. 2, we consider a simple two-level system which exhibits a change in transition frequency when a Stark pulse of amplitude ε is applied. Initially, molecules of velocity v are excited in steady-state by laser light of frequency Ω, thereby placing the transition level in coherent superposition. When the pulse appears, this velocity group is no longer in resonance, but because of its preparation it will begin to radiate, by analogy with NMR,[14] a free induction decay signal. At the same time, a second velocity group v' will be switched into resonance and will alternately absorb and emit radiation. This

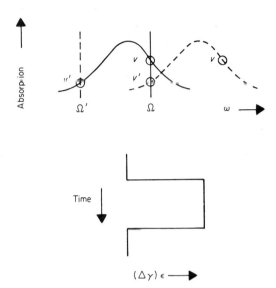

FIGURE 2. Optical switching behaviour of a Doppler-broadened transition when a Stark pulse of amplitude ε is applied. The Stark shift is $(\Delta\gamma)\varepsilon$, the laser frequency is Ω and coherent emission can occur at Ω', for example. The molecular transition frequency is ω; the molecular velocities are v and v'. (from Ref. 16)

optical ringing or nutation effect[12] appears as a damped oscillation. Finally, when the pulse terminates, the group v is suddenly excited and it too begins to nutate, while the second group v' now emits a FID signal. If two pulses are applied, it is possible to observe a photon echo, the optical analogue of a spin echo.[11]

Transient light signals that are emitted by molecules switched out of resonance propagate in the forward direction, because of the preparative step, and are monitored together with the transmitted laser beam by a photodetector. Heterodyne detection is possible, as in the Raman beat[6] and photon echo experiments,[1, 4] since the emission signal is Stark-shifted from the laser frequency; heterodyne detection increases the emission signal amplitude 1000 fold and in the infrared, enhances detection sensitivity.

Most of the experiments discussed here involve an infrared vibration-rotation transition of $^{13}CH_3F$, a molecule that exhibits a first order Stark shift. The transition is a fundamental v_3 band $R(4)$ line, $(J, K) = (4, 3) \rightarrow (5, 3)$ that overlaps the $P(32)$ CO_2 laser line at 1035.474 cm^{-1} (9.66 microns). The CO_2 laser, which is described elsewhere,[17] is free running and possesses high amplitude and frequency stability. The output is ~ 1 watt and the beam is expanded by a Galilean telescope to ~ 1 cm diameter to lengthen the molecule–optical interaction time. Furthermore, the light is linearly polarized, permitting $\Delta M = 0$ selection rules when its orientation is parallel to the dc Stark field or $\Delta M = \pm 1$ transitions when it is perpendicular to the Stark field.

OPTICAL NUTATION

The optical nutation effect is illustrated in Fig. 3 and can be described formally by a solution of the coupled Maxwell-Schrödinger equations, as is true for all of the coherent transients discussed in this article. The nutation derivation[18, 19] has been presented previously, in various degrees of approximation and will not be reproduced here. We only note that the oscillation or nutation frequency of the transition a↔b is given by $2\alpha = \mu_{ab}E_0/\hbar$. It is evident that the dipole matrix element μ_{ab} can be determined experimentally and rather easily from the nutation frequency 2α and the optical field strength E_0.

Coherent optical transients may be observed either under adiabatic or nonadiabatic switching conditions. In Fig. 3, the nonadiabitc condition is realized because the rate at which the transition frequency ω is changed satisfies $d\omega/dt \gg \alpha^2$ where $d\omega/dt \sim 10^3\,\text{MHz}^2$ and $\alpha^2 \sim 1\,\text{MHz}^2$. In the other regime, $d\omega/dt \ll \alpha^2$, adiabatic fast passage experiments have been performed[7, 8] leading to a population inversion and a measurement of T_1 for the excited state.

Stark-switching can be effective only if the frequency shift is larger than the homogeneous width. In the present example of a dilute molecular CH_3F gas sample at $\sim 1\,\text{mTorr}$ pressure, the homogeneous line-width is quite narrow, $\sim 100\,\text{kHz}$, so that rather modest Stark fields of a few tens of volts/cm are adequate.

The decay behaviour of the nutation signal depends in general on several dephasing processes. It includes:

(1) the decay time T_1 arising from the molecular time of flight across the laser beam and from molecular collisions that change the population (diagonal density matrix elements) of the transition levels;

(2) the decay time T_2 due to molecular collisions that disturb the phase of the off-diagonal density matrix element;

(3) spatial inhomogeneities in the optical field profile.

However, the decay time T_1 can be extracted simply from the nutation experiment suggested by Fig. 3.[19] Notice that during steady-state excitation the laser burns a hole within the Gaussian velocity distribution. When a Stark pulse of width τ appears, the partially saturated velocity group v is suddenly switched out of optical resonance with the laser frequency and emits a short-lived FID signal. Simultaneously, a second velocity group v', which need not concern us, is switched into resonance giving the first optical nutation pattern of Fig. 3. When the pulse terminates, the initial velocity

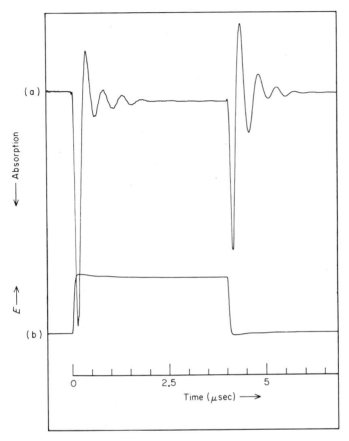

FIGURE 3. (a) Optical nutation in $^{13}CH_3F$ following a Stark pulse of amplitude $\varepsilon = 35$ V^{-1} cm. (from Ref. 1)

group v is switched back into resonance and generates the second nutation pattern. The amplitude of the *delayed nutation*, which is of interest, depends on the population of the group v at the end of the pulse and thus on the extent to which the hole has been filled during the pulse interval. (Note that the group v' simultaneously emits a FID signal which is unimportant because it is short lived). The dependence of the delayed nutation amplitude on pulse width τ can be shown[19] to be proportional to the population difference

$$w(\tau) = w_0 - [w_0 - w(0)] e^{-\tau/T_1} \qquad (1)$$

where in terms of the diagonal elements of the transition levels a and b $w(\tau) = \rho_{aa}(\tau) - \rho_{bb}(\tau)$ at the end of the pulse, $w(0)$ is the population difference at time zero preceding the pulse during steady-state excitation, and w_0 is the population difference in the absence of radiation. We see that the second nutation pattern of Fig. 3 will grow with increasing pulse width τ, according to equation (1), and that T_1 may be readily obtained from the envelope function.

The above discussion presumes that the saturation hole is filled dominantly by molecules that have suffered a change in quantum state rather than by a molecular velocity diffusion mechanism. The latter process, as will be seen from the photon echo measurements, is indeed a much slower effect.

FREE INDUCTION DECAY AND INTERFERENCE PULSES

Another coherent transient effect that can be observed easily in the optical region by Stark switching is free induction decay (FID).[5] The FID effect is shown in Fig. 4 for a nondegenerate transition of NH_2D. Initially, the molecules exist in two-level superposition states during steady-state excitation. The dipoles induced in the sample during this preparative stage then radiate a coherent beam of light in the forward direction when a Stark field is

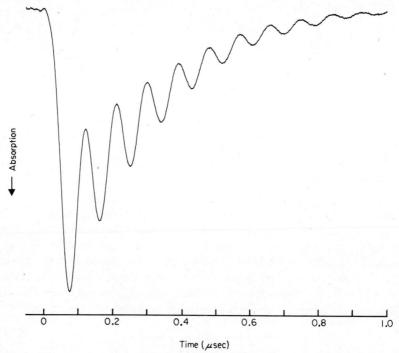

FIGURE 4. Optical free induction decay in NH_2D following a step function Stark pulse. The beat frequency is the Stark shift and the slowly varying background is a nutation signal, more clearly shown in Fig. 3. (from Ref. 5)

suddenly applied. The heterodyne beat signal displayed in Fig. 4 is due to the Stark shift, and the rapid decay results from a destructive interference of the different frequencies emitted, corresponding to the various velocity packets excited in the steady-state preparation. A solution[5,18] of the Maxwell-Bloch equations for this case yields a free induction field that decays as

$$E \propto \exp[-\{1 + (1 + 4\alpha^2 T_1 T_2)^{\frac{1}{2}}\}t/T_2] \qquad (2)$$

where $\alpha = \mu_{ab} E_0/2\hbar$. Equation (2) is confirmed by experiment[5] for times $t \gg T_2^*$ where $1/T_2^*$ is a measure of the Doppler width; the subtle behaviour near the time origin has also been explored[10] but need not be examined here.

The interference effect just mentioned cuts short the emission process but the information is not lost, and in fact this dephasing effect can be reversed by a second pulse which gives rise to the photon echo discussed in the next section.

In the above discussion, we have considered coherent emission in a collection of two-level systems that are not degenerate. A new effect[10] arises, however, when the transition levels are degenerate during the steady-state preparative stage. The emission that results after a step-function Stark field is applied is no longer a simple decay but instead appears as a train of sharp pulses regularly spaced in time due to a repetitive interference. The effect is shown in Fig. 5 for $^{13}CH_3F$ where the selection rule is $\Delta M = 0$. This situation arises because an entire set of infrared transitions within the Stark split manifold are initially prepared. This emission, which beats with the laser, produces a heterodyne beat spectrum consisting of a set of regularly spaced frequencies that is the Fourier transform of the slowly decaying pulse

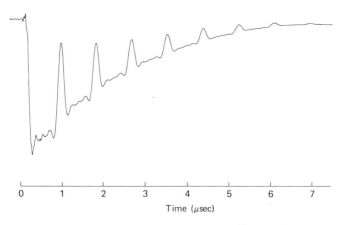

FIGURE 5. Free induction decay interference pulses in $^{13}CH_3F$. (from Ref. 10)

progression observed. By means of a spectrum analyser, the time behaviour of Fig. 5 can be Fourier transformed into the spectrum given in Fig. 6.

We thus demonstrate what is the optical analogue of the well known NMR method of high resolution pulse Fourier spectroscopy. In this simple example, the regular frequency interval between spectral lines is due to the first order Stark shift, but with other molecules, hyperfine structure and other spectral features might be investigated where very high resolution is required. In Fig. 6, the line width is ~ 0.5 MHz, considerably narrower than the 66 MHz Doppler width. The usual advantages of pulse Fourier spectroscopy apply, namely:

(1) an enhanced signal-to-noise ratio compared with steady-state techniques, because of averaging repetitively produced signals;
(2) rapid data acquisition;
(3) the possibility of monitoring *spectra* as a function of decay time in a variety of coherent transient experiments.

These interference pulses superficially resemble mode locking of a laser because several evenly spaced frequency components are involved. However, in this case the regular frequency spacing is inherent in the transitions involved and no interaction between them is needed. The detailed behaviour of the pulse train agrees, in fact, with an FID theory[10] that assumes the transitions to be uncoupled. A simplified version of this theory for $\Delta M = 0$ selection rules shows that the photodetector will monitor cross terms in the optical intensity of the form

$$E^2 \propto \frac{\sin(9\delta\varepsilon t/2)}{\sin(\delta\varepsilon t/2)} \quad (3)$$

where $\delta\varepsilon$ is the frequency interval between two neighbouring lines of Fig. 6 due to a dc Stark field ε and the number 9 signifies the number of transitions involved. Equation (3) shows the characteristics displayed in Fig. 5, i.e., a pulse interval given by $T = 2\pi/\delta\varepsilon$ and a pulse width $\Delta t = 2\pi/9\delta\varepsilon = T/9$ so that their ratio $T/\Delta T = 9$ is just the number of transitions contributing. The more rigorous theory[10] gives $T/\Delta T = 6.25$. The time scale becomes compressed as either the number of transitions, the Stark tuning rate δ or the Stark field ε increases. It is tempting to consider the rise of higher Stark fields. At the present time, pulse widths of ~ 10 nsec have been realized with a modest Stark field of ~ 350 V/cm. At a field of $\sim 100,000$ V/cm, coherent infrared pulses of ~ 50 psec might be achieved.

PHOTON ECHOES AND MOLECULAR COLLISIONS

A unique application of coherent optical transient phenomena is that specific collision mechanisms can be isolated and examined separately. This

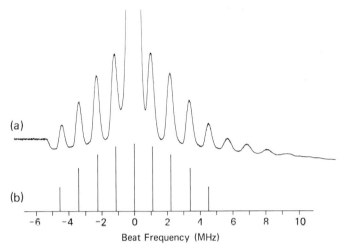

FIGURE 6. $^{13}CH_3F$ heterodyne beat spectrum of FID pulses shown in Fig. 5. A Hewlett-Packard 8553B Spectrum Analyser was used. (a) Observed spectrum. (b) Predicted spectrum. The vertical scales are linear. In (a), beats are not seen below -5 MHz due to instrumental limitations and the beats beyond $+5$ MHz appear to be spurious. (from Ref. 10)

contrasts sharply with the traditional steady-state linewidth measurement which reflects contributions from all relaxation processes. As an example, we shall concentrate on the use of photon echoes in studying elastic collisions that change molecular velocity but not quantum state.[4,19]

To produce photon echoes in a molecular gas, one first subjects the sample to a radiation pulse by switching the molecules into resonance with a Stark pulse. This produces an array of dipoles that are initially in phase. These dipoles radiate and begin to dephase due to their motion. The relative dephasing of a molecule moving with velocity $\mathbf{v}$ will be $\mathbf{k}\cdot\mathbf{v}t$ where $\mathbf{k}$ is the radiation propagation vector and t the time elapsed after the applied pulse.

Another pulse applied to the sample at time τ (see Fig. 7) can produce a phase change of $-2\mathbf{k}\cdot\mathbf{v}\tau$ in molecules moving with velocity $\mathbf{v}$ so that the relative phase of such molecules at time τ will be $\mathbf{k}\cdot\mathbf{v}\tau - 2\mathbf{k}\cdot\mathbf{v}\tau = -\mathbf{k}\cdot\mathbf{v}\tau$. Since the phase of each molecule still increases an amount $\mathbf{k}\cdot\mathbf{v}\Delta t$ in a time Δt, one easily sees that at time $t = 2\tau$ the molecules are all in phase once again and an "echo" pulse is emitted. The photon echo amplitude will serve to monitor all relaxation processes that interfere with this dephasing-rephasing cycle and is a sensitive probe of collision effects in gases.

In particular, if some of the molecules undergo a change in velocity $\Delta\mathbf{v}$ due to a collision, the echo electric field amplitude will be of the form

$$E \sim \langle e^{i\mathbf{k}\cdot\Delta\mathbf{v}\tau} \rangle \qquad (4)$$

where $\langle\ \rangle$ is a collision average. If $k\Delta v\tau \gg 1$, any collision produces destruc-

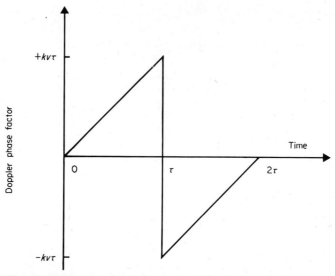

FIGURE 7. The relative Doppler phase of molecules with velocity v during a two-pulse echo sequence. The molecules are all in phase because of a $\pi/2$ pulse at $t = 0$. The phase then advances but reverses sign at $t = \tau$ by application of a π pulse. At $t = 2\tau$, the molecular dipoles have all dephased again and an echo signal is produced. (from Ref. 19)

tive interference so that the only term which survives in equation (4) will be the one in which no collision occurs during the time $t = 2\tau$. Since the associated probability is $e^{-\Gamma t}$, where Γ is the rate of elastic collisions, one finds that for

$$k\Delta v\tau > 1; \quad E \sim e^{-\Gamma t}. \tag{5}$$

On the other hand, if $k\Delta v\tau \ll 1$, each collision produces only a small phase change such that

$$E \sim 1 - \frac{k^2}{2} \langle (\Delta \vec{\mathbf{v}})^2 \rangle \tau^2 \tag{6}$$

where we have assumed for simplicity that $\langle \Delta v \rangle = 0$. The quantity $\langle (\Delta v)^2 \rangle$ will be equal to (number of collisions in time $t = \Gamma t$) $\times$ [$\langle \Delta v^2 \rangle$ for one collision $= \Delta u^2/2$] so that for

$$k\Delta v\tau < 1; \quad E \sim \exp[-\Gamma t^3 (k\Delta u)^2/16]. \tag{7}$$

Equations (5) and (7) agree, apart from numerical factors, with a more precise derivation, given in Refs 4 and 19, that begins with the Boltzmann transport equation

$$\begin{aligned}\partial \rho_{mn}(\mathbf{R},\mathbf{v},t)/\partial t = &-\mathbf{v} \cdot \partial \rho_{mn}(\mathbf{R},\mathbf{v},t) - \rho_{mn}(\mathbf{R},\mathbf{v},t)/T_1 - \Gamma \rho_{mn}(\mathbf{R},\mathbf{v},t) \\ &+ \int d^3v' W(\mathbf{v}' - \mathbf{v})\rho_{mn}(\mathbf{R},\mathbf{v}',t) + [H(\mathbf{r},\mathbf{R},t), \rho(\mathbf{R},\mathbf{v},t)]_{mn}/i\hbar,\end{aligned} \tag{8}$$

where $\rho_{mn}(\mathbf{R}, \mathbf{v}, t)$ is the phase-space distribution function for the mnth density-matrix element. The off-diagonal density-matrix elements, which depend on molecular position $\mathbf{R}$ and velocity $\mathbf{v}$, give rise to a sample polarization and to an echo radiation pattern derivable from Maxwell's equations. The first term on the right-hand side of equation (8) denotes the spatial behaviour or convective flow, the second term explicitly expresses the population decay through the constant T_1, the third and fourth terms result from velocity-changing collisions that respectively decrease or increase $\rho_{mn}(\mathbf{R}, \mathbf{v}, t)$, and the last term is the quantum-mechanical commutator appearing in Schrödinger's equation. Thus, in the absence of the last term, equation (8) reduces to the linearized Boltzmann equation. Note that $H(\mathbf{r}, \mathbf{R}, t)$ is the Hamiltonian for the system excluding collisions, $W(\mathbf{v}' - \mathbf{v})$ is the probability density per unit time for a collision that takes a molecule from velocity $\mathbf{v}'$ to $\mathbf{v}$, and Γ is the rate constant given by $\Gamma = \int d^3 v W(\mathbf{v}' - \mathbf{v})$ for velocity-changing collisions. Finally, there is no term in equation (8) representing "phase-interrupting collisions." Its absence is consistent with our assumption of a state-independent collisional interaction.

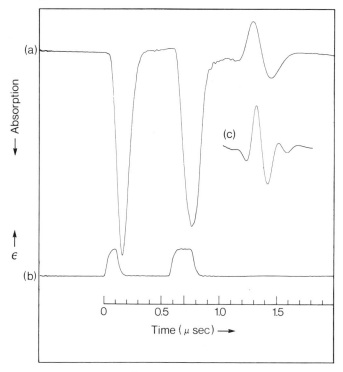

FIGURE 8. Two-pulse photon echo in $^{13}CH_3F$. (a) Optical response, (b) Stark-pulse sequence with $\varepsilon = 35$ V^{-1} cm. (from Ref. 1)

To solve equation (8), we assume the Keilson and Storer[20] "weak" collision kernel

$$W(\mathbf{v}' - \mathbf{v}) = \Gamma[\pi(\Delta u)^2]^{3/2} \exp[-(\mathbf{v} - \alpha\mathbf{v}')^2/(\Delta u)^2]. \tag{9}$$

The velocity jump Δu is related to the most probable speed u of the thermal-equilibrium distribution by

$$(\Delta u)^2 = (1 - \alpha^2)u^2 \approx 2(1 - \alpha)u^2, \tag{10}$$

and α is a constant very close to but less than 1. Inserting equation (9) into equation (8) leads to an analytic solution for the echo amplitude under the following assumptions:

(1) the molecular sample is subject to resonant infrared pulses, a $\pi/2$ pulse at $t = 0$ and a π pulse at $t = \tau$, which are sufficiently brief that collisions and spatial terms in equation (8) can be neglected, leaving only the Schrödinger equation. Following a pulse, the off-resonant condition prevails and all terms in equation (8) are needed to describe relaxation and echo formation;

(2) the average velocity of the sample decays slowly, i.e., $\Gamma(1 - \alpha)\tau \ll 1$, where $\Gamma(1 - \alpha)$ is the effective decay rate of the average velocity. The collision rate Γ is taken to be much less than the Doppler width, $\Gamma/ku \ll 1$;

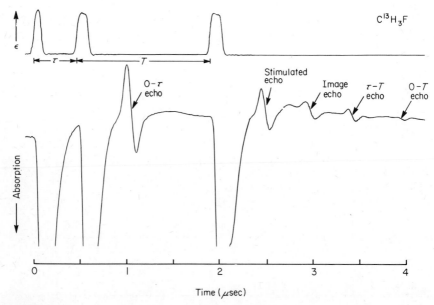

FIGURE 9. Three-pulse photon echo in $^{13}CH_3F$. *Top:* Stark-pulse sequence. *Bottom:* Optical response.

(3) the initial velocity distribution at $t = 0$ is the thermal equilibrium value $(1/\pi u^2)^{1/2} \exp(-v^2/u^2)$.

We thus obtain the normalized echo amplitude E, evaluated at $t = 2\tau$:

$$E(t = 2\tau) = \exp[-t/T_1 - \Gamma t + (4\Gamma/k\Delta u)\int_0^{kt\Delta u/4} e^{-x^2} dx]. \quad (11)$$

This has the limiting form for short times, $kt\Delta u < 1$,

$$E(t = 2\tau) = \exp[-t/T_1 - \Gamma k^2(\Delta u)^2 t^3/48], \quad (12)$$

and for long times, $kt\Delta u > 1$,

$$E(t = 2\tau) = \exp[-t/T_1 - \Gamma t + (2\Gamma/k\Delta u)\pi^{\frac{1}{2}}]. \quad (13)$$

We see the general expression (11) reduces to the limiting cases, equations (12) and (13), in agreement with our earlier estimate, equations (7) and (5).

An example of a two pulse echo[11] is presented in Fig. 8, a three pulse or stimulated echo[11] in Fig. 9, and multiple pulse or Carr-Purcell echoes[13] in Fig. 10. Each technique provides unique information about molecular collisions. The stimulated echo, for example, allows one to measure the decay time T_1, and we shall not consider it further because T_1 can be measured more easily with the delayed nutation technique described earlier. The Carr-Purcell echoes allow one to measure the dephasing time T_2 without the complication of elastic collisions, where ideally the pulse sequence is a $\pi/2$ pulse followed by a succession of π pulses as in Fig. 9. This T_2 measurement is not only sensitive to inelastic collisions and to the molecular transit time through the laser beam (the T_1 part) but also to molecular collisions that influence the phase of the off-diagonal density matrix element. This implies

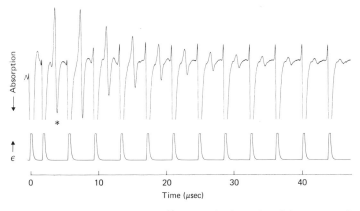

FIGURE 10. Optical Carr-Purcell echoes in $^{13}CH_3F$. The first echo of the sequence is marked by an asterisk. Conditions are the same as in Fig. 11. (from Ref. 4)

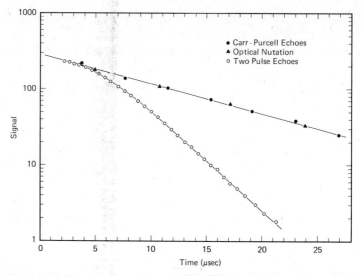

FIGURE 11. Decay curves of optical coherent transients in $^{13}CH_3F$. The pressure is 0.32 mTorr, the laser beam diameter matches a 1.3 cm Stark spacing, the Stark pulses are 40 V^{-1} cm and a bias field of 80 V^{-1} cm is added to remove any complication of level degeneracy. The expanded laser beam's power density is ~ 350 mW cm^{-2}. The time axis is the total elapsed time t for each experiment so that $t = 2\tau$ in a two-pulse echo or $t = 2n\tau$ in an n-pulse echo train. (from Ref. 4)

that a Carr-Purcell measurement will reflect the importance of phase-interrupting collisions. In Fig. 11, we see that the T_1 decay from a delayed nutation measurement coincides with a T_2 decay from a Carr-Purcell experiment so that phase interrupting collisions are of little importance for this vibration-rotation transition of $^{13}CH_3F$. We note finally that the Carr-Purcell echo amplitude expression for an n pulse sequence

$$E_n(t = 2n\tau) = \exp[-t/T_1 - \Gamma t + 4n\Gamma/k\Delta u) \int_0^{k\tau\Delta u/2} e^{-x^2} dx]$$

follows by taking equation (11) to the nth power where we have set $T_2 = T_1$. For short pulse separations, $k\tau\Delta u \ll 1$,

$$E_n(t = 2n\tau) = \exp[-t/T_1 - \tfrac{1}{12}\Gamma(k\tau\Delta u)^2 t], \tag{15}$$

and the second term on the right due to elastic collisions can be made arbitrarily small by reducing the interval τ so that the T_1 decay dominates.

We return now to the two-pulse echo where its decay behaviour is shown in the bottom curve of Fig. 11. The decay is not a simple exponential but shows an unusual time dependence that departs first slowly from the T_1 behaviour (upper curve) and then more rapidly at longer times. Closer examination shows that there are two limiting time regimes of the form

$$e^{-Kt^3} \text{ for short times,} \qquad (16a)$$

$$e^{-\Gamma t} \text{ for long times.} \qquad (16b)$$

The long term linear decay is clearly evident over two decades of signal amplitude but the cubic time dependence is better seen on a t^3 time axis (not shown here). This observed functionality in time is precisely what is predicted by equations (12) and (13). The agreement constitutes primary evidence that elastic molecular collisions involving small changes in velocity play a crucial role in photon echo measurements of molecular gases. The departure of the two-pulse echo from the T_1 decay is due solely to the velocity-changing collision mechanism. That the T_1 decay is independent of elastic collisions is confirmed by other measurements involving the Raman beat effect and is described elsewhere.[6,19]

From the observed time dependence, equation (16), we obtain the elastic $^{13}CH_3F-^{13}CH_3F$ collision parameters

$$\sigma = 430 \text{ Å}^2$$
$$\Delta u = 85 \text{ cm/sec.} \qquad (17)$$

where the total elastic cross section is defined by $\sigma = \Gamma/(Nu\sqrt{2})$ and N is the molecular number density.

The characteristic velocity jump per collision Δu is seen to be quite small, $\sim 0.2\%$ of thermal velocity, and thus, velocity jumps are closely clustered about the initial velocity, justifying our use of a weak collision model. Furthermore, it follows that an arbitrary initial velocity approaches a thermal equilibrium distribution as $\exp(-\beta t)$ in a time $\beta^{-1} = [\Gamma(1-\alpha)]^{-1} = 5$ sec which contrasts greatly with the time scale of an echo experiment $\Gamma^{-1} \sim 14$ μsec at 1 mTorr pressure.

The pressure dependent part of T_1 represents still other collision processes, namely, those that involve population changes and hence a change in quantum state, the most conspicuous being jumps in molecular rotation state (J) and orientation (M). Recent optical double resonance studies of $^{13}CH_3F$ have indicated efficient transfer among neighbouring M states, due to the permanent dipole–dipole interaction.[21] This interaction presumably plays the dominant role in elastic collisions as well. Our results are summarized in Table 1.

TABLE 1. $^{13}CH_3F-^{13}CH_3F$ Collision Parameters

	Cross section (Å²)	
Velocity-changing collisions	430	; $\Delta u = 85$ cm^{-1}/sec
Rotation ($\Delta J = \pm 1, \pm 2, \ldots$)	400	
Reorientation ($\Delta M = \pm 1, \pm 2, \ldots$)	70	
Phase-interrupting collisions	~ 0	

REFERENCES

1. Brewer, R. G., and Shoemaker, R. L. (1971), *Phys. Rev. Lett.* **27**, 631.
2. Kurnit, N. A., Abella, I. D., and Hartmann, S. R., (1964), *Phys. Rev. Lett.* **13**, 567; ibid (1966), *Phys. Rev.*, **141**, 391.
3. Abella, I. D., Compaan, A., and Lambert, L. Q., (1974), in *"Laser Spectroscopy,"* Brewer, R. G., and Mooradian, A. eds, Plenum, New York, p. 457 and references therein.
4. Schmidt, J., Berman, P. R., and Brewer, R. G. (1973), *Phys. Rev. Lett.*, **31**, 1103.
5. Brewer, R. G., and Shoemaker, R. L. (1972), *Phys. Rev.* **A6**, 2001.
6. Shoemaker, R. L., and Brewer, R. G., (1972), *Phys. Rev. Lett.*, **28**, 1430; Brewer, R. G., and Hahn, E. L., (1973), *Phys. Rev.* **A8**, 464; Brewer, R. G., and Hahn, E. L., (1975), *Phys. Rev.* **A11**, 1641.
7. Loy, M. (1974), *Phys. Rev. Lett.*, **32**, 814.
8. Levy, J. M., and Brewer, R. G. (unpublished).
9. Schmidt, J., and Brewer, R. G. (unpublished).
10. Foster, K. L., Stenholm, S., and Brewer, R. G. (1974), *Phys. Rev.*, **A10**, 2318.
11. Hahn, E. L. (1950), *Phys. Rev.*, **80**, 580.
12. Torrey, H. C. (1949), *Phys. Rev.*, **76**, 1059.
13. Carr, H. Y. and Purcell, E. M. (1954), *Phys. Rev.*, **94**, 630.
14. Hahn, E. L. (1950), *Phys. Rev.*, **77**, 297.
15. Abragam, A. (1961), The Principles of Nuclear Magnetism, Oxford University Press. 63.
16. Brewer, R. G. (1972), *Science* **178**, 247.
17. Freed, C. (1969); *IEEE J. Quantum Electron.*, **4**, 404; ibid (1967), **3**, 203.
18. Hopf, F. A., Shea, R. F., and Scully, M. O. (1973), *Phys. Rev.*, **A7**, 2105.
19. Berman, P. R., Levy, J. M., and Brewer, R. G. (1975), *Phys. Rev.*, **A11**, 1668.
20. Keilson, J., and Storer, J. E. (1952), *Q. Appl. Math.*, **10**, 243.
21. Brewer, R. G., Shoemaker, R. L., and Stenholm, S. (1974), *Phys. Rev. Lett.*, **33**, 63; (1974), *Phys. Rev.*, **A10**, 2037.

Chapter 8

Nonlinear Optical Techniques for Generation of VUV and Soft X-ray Radiation*

by S. E. Harris, Stanford University, Stanford, California, U.S.A.

As we attempt to construct lasers at progressively shorter wavelengths, the rapidly decreasing spontaneous emission time makes it increasingly difficult to obtain an inversion. Several years ago Dr. J. F. Young, several graduate students, and myself began work on using nonlinear optical processes for frequency conversion of high-power infrared and visible laser radiation into the vacuum ultraviolet and perhaps into the soft X-ray spectral region. The principal new ideas involved the use of resonances of metal vapours and other atomic species, and also the use of phase-matching techniques which allow great increases in the efficiency of nonlinear processes. Over the past several years our experiments have been quite successful, and it is now possible quite conveniently to obtain tunable radiation over much of the vacuum ultraviolet spectral region, and there is some promise of the extension of these techniques to the near soft X-ray region.

In this talk I will first discuss the fundamentals and the experimental results of these frequency tripling and frequency multiplication techniques. I will then discuss their application to early holographic experiments. Finally, I will describe a new type of nonlinear process employing a van der Waals interaction during collision. This latter process allows us to create large cross sections for inelastic collisions between non-resonant species, and may

---

* The work reviewed in this paper was supported by the University of California and the Atomic Energy Commission under Contract Numbers LLL 8532009 and LASL XP3-06745-1, by the National Aeronautics and Space Administration under Contract Number NGL-05-020-103, by the Office of Naval Research under Contract Number N00014-67-A-0112-0036, and by the Joint Services Electronics Programme through the Office of Naval Research.

have quite interesting applications to the problem of generating short wavelengths. Before beginning, I would like to note that I will be summarizing the work of Dr. J. F. Young, Dr. A. H. Kung, and graduate students G. W. Bekkers, G. C. Bjorklund, D. M. Bloom, D. B. Lidow, and E. A. Stappaerts.

I will first describe experiments which involve frequency tripling of radiation at 1.06 μm. Our work on frequency tripling began at this wavelength because of the high peak powers available from 1.06 μm Nd:YAG and glass laser systems and because of the interest in converting this wavelength to 3547 Å for use in fusion applications.

There are basically three general ideas that are applicable to the frequency conversion work.[1-4] The first of these is the use of resonances to enhance greatly optical nonlinearities. For frequency tripling of 1.06 μm, alkali metal vapours are particularly appropriate. As seen in Fig. 1, the resonances of, for instance, Rb, intersperse the fundamental frequency ω_0, $2\omega_0$, and $3\omega_0$. As we move further into the vacuum ultraviolet, more appropriate species change from the alkali metal vapours that are applicable to the visible region to such metal vapours as Cd and Hg. Still further in the vacuum ultraviolet, the inert gases become the most appropriate media for generation. At still shorter wavelengths, as soft X-ray wavelengths are approached, ionized media may provide the most appropriate species. In general, by making use of resonances such as shown in Fig. 1, it is possible to increase the nonlinearity by many orders of magnitude. For instance, the nonlinearity for

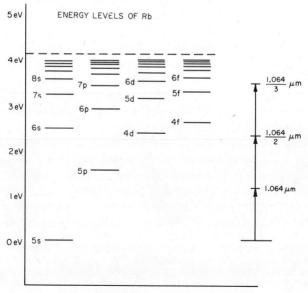

FIGURE 1. Schematic of energy levels of Rb.

frequency tripling of 1.06 μm in Rb is about 10^5 times larger than that for frequency tripling in He.

So the first idea is to make use of resonances. Now the question is which resonances should be used to obtain the largest effect? If one attempts to tune right up to an allowed transition, for instance the 5s–5p resonance of Rb, the optical nonlinearity increases, but two opposing effects take place.[4] First the coherence length becomes very short. In fact, if phase matching is not used, the increase in optical nonlinearity is exactly offset by the decrease in coherence length. If one approaches the resonance line still closer, absorption also becomes significant. If $3\omega_0$ approaches a resonance the same effects take place. We thus conclude that the most desirable resonance to approach is that of a non-allowed transition to ground, for example, the 5s–6s or the 5s–4d in Rb.[5] The closer that the sum of two incident photons is to the frequency of this transition the better off one will be. Depending on the incident power density, the limit of approach to this transition is that of two-photon absorption.

In very physical terms, as we apply an input frequency to the atomic system the electron cloud begins vibrating at the incident frequency ω_0. In the next order of perturbation a pulsation or vibration at $2\omega_0$, primarily associated with the 5s to 4d transition in Rb, appears. Since these transitions are symmetric, no radiation at $2\omega_0$ occurs. Finally, in the next order of perturbation the electron cloud oscillates with a component at $3\omega_0$ where an allowed dipole moment exists and radiation occurs.

The second key idea is that of phase matching. We use the dispersion of one species exactly to cancel that of a second species. The alkali metals are negatively dispersive for any frequency above their resonance line. As seen in Fig. 2, the resonance line of Rb occurs at 7800 Å. For any frequency above

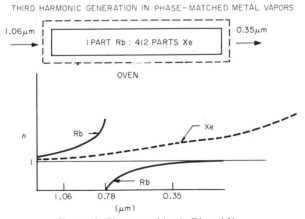

FIGURE 2. Phase matching in Rb and Xe.

this the refractive index is less than one and the phase velocity is greater than that of light in vacuum. This is a result of the very large oscillator strength associated with the resonance transition of these metals. We offset this negative dispersion by mixing with a positively dispersive species such as Xe. For a particular choice of input frequencies there exists a certain ratio of species such that the phase velocity of the generated third harmonic is identical to that of the phase velocity of the incident fundamental. For Rb and Xe this ratio is 412 parts Xe to 1 part Rb. It may also be possible to accomplish vapour–metal vapour phase matching with a greatly reduced ratio. Recent work at Stanford University by D. M. Bloom and J. F. Young shows that this technique has promise.

Figure 3 illustrates the last of the key ideas that I will talk about. This one is a little bit more difficult to explain quickly. It concerns the relationship between focusing and the dispersion of the media. When we examine a Gaussian beam as it passes through a focus one finds a phase slip of π radians, referenced to the phase of a plane wave in the same medium. It is only possible to compensate for the phase slip of the third harmonic wave as compared to that of the fundamental wave in a medium which is negatively dispersive. As seen in Fig. 3 this is quite important experimentally. Thus for example if one takes a ruby laser and focuses its output tightly into a positively dispersive gas such as Xe, then irrespective of the nonlinearity of the Xe, the output signal will be very small. This is a result of cancellation at the output of the focus of harmonics generated at the input of the focus. Figure 3 shows a typical curve of third harmonic output as a function of vapour pressure in a negatively dispersive medium as opposed to a positively dispersive medium.

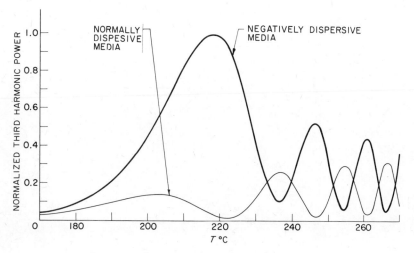

FIGURE 3. Effect of focusing on harmonic generation.

Even though the media are assumed to have the same nonlinearity, the output signals are substantially different. The tighter the focus, the more severe will be the cancellation in the normally dispersive media.

Figure 4 illustrates the experimental techniques which we now use for obtaining many coherence lengths of phase-matched metal vapours. We make use of a concentric heat pipe oven.[6, 7] The inner oven contains a mixture of the metal vapour and the inert gas. We require temperature control typically of the order of a few tenths of 1°C for 50 cm. This is obtained by means of an outer heat pipe. The pressure of the Ar in this outer pipe, in effect, determines the boiling point of the Na which in turn holds the temperature of the inner heat pipe constant. These techniques work well, and have allowed us to obtain over 100 coherence lengths of Na in about 50 cm.

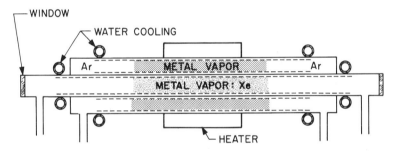

FIGURE 4. Schematic of concentric heat pipe oven.

In particular, no metal vapour is in contact with the end windows. There are some problems. The most severe of these occurs at the boundary zone between the hot Na and the cold inert gas. At pressures much above 5 torr of Na, condensation or fogging in this boundary layer leads to severe beam distortion. Rather than attempt to engineer around this problem, in coming months we will attempt to replace the inert gas by a second metal vapour, which will allow a greatly reduced ratio and probably eliminate this problem.

Figure 5 shows conversion efficiency versus power density for frequency tripling in Rb. At an incident power density of about 10^{10} W/cm^2 an energy conversion of 10% from 1.06 μm to 0.35 μm is obtained. I should note that these experiments, performed by D. M. Bloom, G. W. Bekkers, and J. F. Young, make use of a nearly diffraction-limited output from a Nd:YAG laser with a pulsewidth of about 30 psec. It is not clear whether as the pulselength is increased, that it will still be possible to obtain this relatively high efficiency. Firstly, there is the possibility of ground state absorption which will deplete the metal vapour, in turn changing the refractive index of the medium and breaking the phase-matching condition. Secondly, as longer pulses are employed, the possibility of gas breakdown increases and a longer cell with a looser focus may have to be employed.

We next turn to experiments aimed at generation further into the vacuum ultraviolet spectral region. Figure 6 shows a summary of recent experiments. These experiments all make use of the same ideas as we have already discussed. We note that the Cd:Ar system shows particular promise of generating 1773 Å at quite high efficiency. A second interesting system is the mixture of Xe:Ar which allows generation to 1182 Å at an efficiency between 1% and 3%. Though this system looks simple because it involves a mixture of only two inert gases, I should note that for reasons which we do not quite understand it is quite difficult to mix these gases with sufficient homogeneity. For consistent results a circulating flow cell is required. The shortest wavelength which we have obtained to date is 887 Å.

In an interesting set of recent experiments, Dr. A. H. Kung has demonstra-

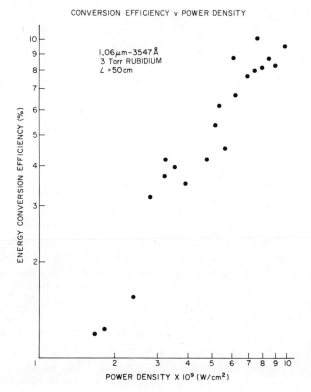

FIGURE 5. Conversion efficiency versus power density for phase-matched Rb vapour.

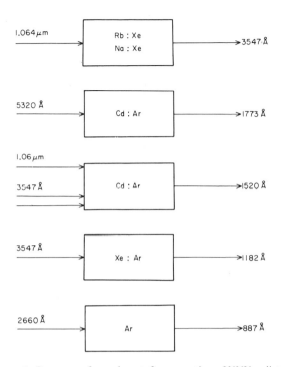

FIGURE 6. Summary of experiments for generation of VUV radiation.

ted tunable picosecond radiation across much of the VUV spectral region.[8] As shown in Fig. 7, his set-up employs a picosecond parametric generator employing two ADP crystals. The first crystal acts as a source of parametric fluorescence or spontaneous emission. This emission passes through a 75 cm long confocal stop and is then amplified in a second parametric ADP amplifier. In this way, pulses which are tunable throughout the entire visible spectral region are obtainable at an efficiency between 5% and 10% of the incident 2660 Å pumping radiation.

In Dr. Kung's experiments this tunable picosecond time-scale radiation is tightly focused into a cell of Xe. The signal and idler outputs of the parametric generator are either frequency tripled or mixed with the pumping

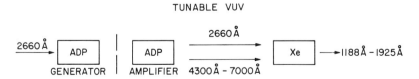

FIGURE 7. Generation of tunable picosecond time scale VUV radiation.

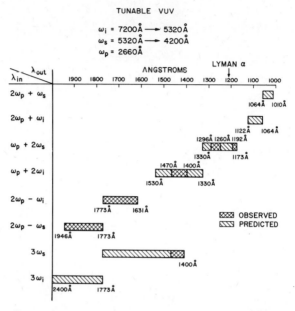

FIGURE 8. Experimental results—tunable VUV generation.

radiation at 2660 Å. Results are shown in Fig. 8. The input frequencies are shown on the left scale, while the generated vacuum ultraviolet radiation is shown on the upper scale. The key point of this figure is that radiation is only obtained in spectral regions where the Xe is negatively dispersive. To the extent that refractive indices are known in the visible and near ultraviolet spectral region, techniques such as this allow their direct measurement in the vacuum ultraviolet spectral region.

Let me now return to an alternative experiment where we make deliberate use of a non-allowed transition.[5, 9, 10] The first experimental work of this type was demonstrated by Hodgson, Sorokin, and Wynne.[5] As shown in Fig. 9, a first laser is tuned so that the frequency sum of two of its photons is exactly equal to that of a non-allowed transition. A second laser is also incident on the cell to produce tunable radiation in the vacuum ultraviolet spectral region. By making exact use of this non-allowed transition, it is possible to get high efficiency of generation at power densities which are about three orders of magnitude lower than that in a typical non-resonant experiment. The highest efficiencies will be obtained when the frequency sum of the three incident photons is near to an allowed transition to ground. In a related experiment of this type employing a CO_2 laser, a parametric oscillator, and Na vapour, we have demonstrated 58% photon conversion from 9.26 μm to 3305 Å.[10] This 58% photon conversion corresponds to a

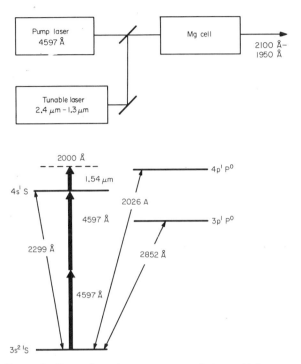

FIGURE 9. Generation using resonant two-photon excitation.

power gain of about 16. Techniques of this type may lead to practical infrared detecting and imaging devices.

One of the motivations for short-wavelength lasers is a desire to take holograms and perhaps do fabrication at dimensions smaller than is possible with optical wavelengths. A number of early proposals for X-ray holography implicitly involved large aperture optics with tolerances of the order of the wavelength employed. Since the fabrication of such optics is not yet possible, we are approaching this area from a somewhat different point of view. As shown in Fig. 10, our approach is to use a far-field Fraunhofer hologram to transfer the information from the object to a photosensitive medium. The readout of this information will then be accomplished with an electron microscope. Image construction could then be obtained with a visible-wavelength laser or by computer techniques. The primary function of the short-wavelength laser will be to obtain increased contrast, as opposed to increased resolution. The hope is that biological samples will have higher contrast for soft X-ray radiation than they do for an incident electron beam. Also, the need for staining techniques would be eliminated. These ideas have been pursued by G. C. Bjorklund and J. F. Young.[11] Using 1182 Å radiation which was produced by harmonic generation, they have constructed

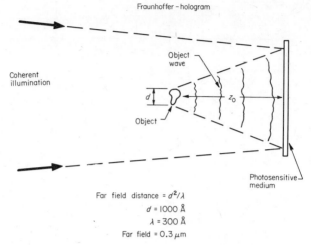

FIGURE 10. Far-field holography at short wavelengths.

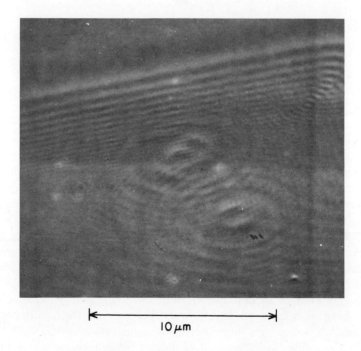

Holograms of two 1.305 μm spheres
(sem)

FIGURE 11. Scanning electron microscope photo of 836 Å spaced fringes.

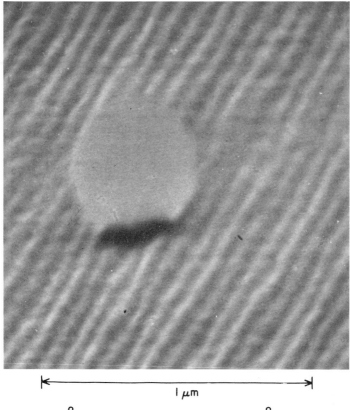

836 Å fringes produced in PMMA by 1182 Å radiation

FIGURE 12. Scanning electron microscope photo of holograms of 1.3 μm diameter particles.

fringes with a spacing as small as 836 Å (Fig. 11). The substrate in these experiments was polymethyl methacrylate. Figure 12 shows scanning electron microscope holograms of 1.3 μm diameter particles taken by the far-field hologram technique.

I would like next to change subject somewhat and describe a new type of process which may have applicability to the problem of generating coherent short wavelength radiation. The type of process I will discuss takes place via a dipole–dipole interaction during the collision of two species.[12-14] As I will describe, it allows increased transition probabilities for higher order nonlinear optical processes and also may allow large cross sections for inelastic collisions between non-resonant species.

As shown in Fig. 13, when an optical wavelength is tuned near to the resonance of some atomic species A, a strong dipole moment at the frequency

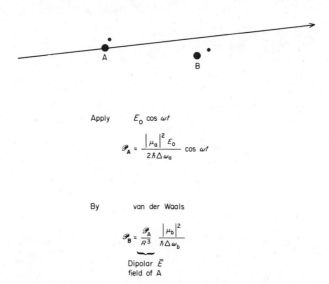

FIGURE 13. Nonlinear optical processes by van der Waals interaction during collision.

of the optical beam or at harmonics of this beam is excited in this species. As atom A passes near to some other species B, the second species is subjected to a near electric field which may be several orders of magnitude larger than the macroscopic electric field applied to the system. This near electric field acts on a B atom in the same manner as would an external field. As an example of the use of this type of process, assume that one were interested in a four-photon absorption process in krypton. As shown in Fig. 14, one could form a mixture of krypton and mercury and choose the incident frequencies such that $\omega_1 = \omega_2 = 3129$ Å, while $\omega_3 = \omega_4 = 7342$ Å. The sum of ω_1 and ω_2 would excite the non-allowed transition in Hg. Via the dipolar interaction, the excitation at frequency $\omega_1 + \omega_2 + \omega_3$ would be transferred to atom B during collision. During this collision process the B atom could accept an additional photon from the electric field, here denoted by ω_4, and complete the four-photon transition. Calculations performed by D. B. Lidow and myself show that at a Hg concentration of 10^{17} atoms/cm^3 that the four-photon transition probability in krypton would be increased by about 10^4.

There is a second and potentially more interesting application of these ideas. As shown in Fig. 15, suppose that it is possible to store atoms in some excited state of a first species, for example, as shown here in the 2s state of He. Suppose also that it were desirable to create a laser in some species B whose

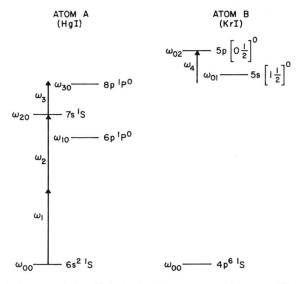

FIGURE 14. Enhancement of multiphoton ionization processes by van der Waals interaction.

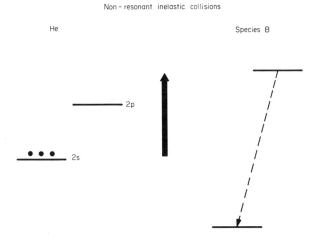

FIGURE 15. Non-resonant inelastic collisions in the presence of strong optical electric fields.

upper state is not resonant with the 2s state of He. The basic idea is to apply an electromagnetic field with a frequency to satisfy the energy defect between the two species. Calculations show that in the presence of quite strong applied electric fields that collision cross sections greater than 10^{-14} cm^2 between the two species can be obtained.

One of the basic problems in constructing short-wavelength lasers is the need to store energy for a reasonable period of time, while at the same time creating an inversion in a time short compared to the spontaneous emission time. The process shown in Fig. 15 may lead to what we term a "switched" collision. Energy could be stored in the first species by optical or electron beam techniques, following which a high-power picosecond pulse at the frequency of the selected energy defect could be applied to the system. As shown in this figure, a rapid inversion in species B could then be obtained.

An alternative application of this technique might be to use the storage in species A as an intermediate step for ionization of innershell electrons in species B. In the Duguay and Rentzepis[15] proposal of several years ago, it was noted that if nanosecond time-scale optical radiation were available in the 300 Å spectral region, this radiation would selectively ionize an innershell electron of Na and create an inversion in the neon-like ion. The idea would be first to store in an intermediate species and using the technique that I have described apply a picosecond pulse to bring the energy of species A to the desired level for the selective ionization.

There is one other potentially interesting though somewhat different application of this general type of idea. That is to make use of a charge exchange collision, where once again the energy defect is supplied by an optical photon. By making use of charge exchange collisions, large energy storage could be obtained, before the triggering optical pulse is applied.

I would like to note again that the work that I have described in this talk involves major contributions from J. F. Young, A. H. Kung, G. W. Bekkers, G. C. Bjorklund, D. M. Bloom, D. B. Lidow, and E. A. Stappaerts.

REFERENCES

1. Harris, S. E., and Miles, R. B. (1971), "Proposed Third Harmonic Generation in Phase Matched Metal Vapors," *Appl. Phys. Lett.* **19**, 385; and Young, J. F., Bjorklund, G. C., Kung, A. H., Miles, R. B., and Harris, S. E. (1971), "Third Harmonic Generation in Phase Matched Rb Vapor," *Phys. Rev. Lett.* **27**, 1551.
2. Kung, A. H., Young, J. F., Bjorklund, G. C., and Harris, S. E. (1972), "Generation of Vacuum Ultraviolet Radiation in Phase Matched Cd Vapor," *Phys. Rev. Lett.*, **29**, 985.
3. Kung, A. H., Young, J. F., and Harris, S. E. (1973), "Generation of 1182 Å Radiation in Phase Matched Mixtures of Inert Gases," *Appl. Phys. Lett.*, **22**, 301.
4. Miles, R. B., and Harris, S. E. (1973), "Optical Third Harmonic Generation in Alkali Metal Vapors," *IEEE J. Quantum Electron.*, **9**, 470.
5. Hodgson, R. T., Sorokin, P. P., and Wynne, J. J. (1974), *Phys. Rev. Lett.*, **32**, 343.
6. Vidal, C. R., and Haller, F. B. (1971), *Rev. Scient. Instrum.* **42**, 1779.

7. Vidal, C. R., and Hessel, M. M. (1972), *J. Appl. Phys.*, **43**, 2776.
8. Kung, A. H. (1974), "Generation of Tunable Picosecond VUV Radiation," *Appl. Phys. Lett.*, **25**, 653.
9. Harris, S. E., and Bloom, D. M. (1974), "Resonantly Two-Photon Pumped Frequency Converter," *Appl. Phys. Lett.*, **24**, 229.
10. Bloom, D. M., Yardley, James T., Young, J. F., and Harris, S. E. (1974), "Infra-red Up-Conversion With Resonantly Two-Photon Pumped Metal Vapors," *Appl. Phys. Lett.*, **24**, 427.
11. Bjorklund, G. C., Harris, S. E., and Young, J. F. (1974), "Vacuum Ultraviolet Holography," *Appl. Phys. Lett.*, **25**, 451.
12. Harris, S. E., and Lidow, D. B. (1974), "Nonlinear Optical Processes by van der Waals Interaction During Collision," *Phys. Rev. Lett.*, **33**, 674.
13. Gudzenko, L. I., and Yakovlenko, S. I. (1972), "Radiative Collisions," *Soviet Phys. JETP*, **35**, 877.
14. Yakovlenko, S. I. (1973), "Ionization of Atoms in Radiative Collisions," *Soviet Phys. JETP*, **37**, 1019.
15. Duguay, M. A., and Rentzepis, P. M. (1967), *Appl. Phys. Lett.*, **10**, 350.

Chapter 9

High Resolution Astronomy between Three Microns and Three Millimetres*

by C. H. Townes, University of California, Berkeley, California, U.S.A.

I have chosen to discuss astronomy of a particular spectral region, with emphasis on new techniques that are developing and the contributions such techniques can make. This spectral region lies between three microns and three millimetres in wavelength—a span corresponding to three decades in wavelength. It is an important and rich region, but so far relatively undeveloped by comparison with the optical or short infrared on one side or the radio and longer wavelength regions on the other. Even in these spectral regions there is room for improvement, perhaps by as much as an order of magnitude, but I consider them to be reasonably well developed. In the visible and in the microwave regions techniques are getting close to some fairly fundamental limits, whereas throughout most of the infrared the room for improvement is more characterized by three orders of magnitude than by one.

To illustrate the state of development on either side of the region I shall primarily discuss, consider the spectroscopy done by Pierre Connes. His Fourier spectroscopy using quite long interferometers has obtained spectacular spectra at very high resolution of some of the more intense astronomical sources. Figure 1 shows such spectra, made by Connes and Michel.[1] The first is a spectrum of lunar light in which one sees atmospheric absorption lines. Next is a spectrum of α Orionis where one can see that the stellar lines are substantially broader than the resolution obtained, which is about 3×10^5. The third spectrum is of another star, R Leonis, which has a rich spectrum containing an enormous amount of information. Characteristically,

*Work supported in part by NASA Grant NGL 05-003-272.

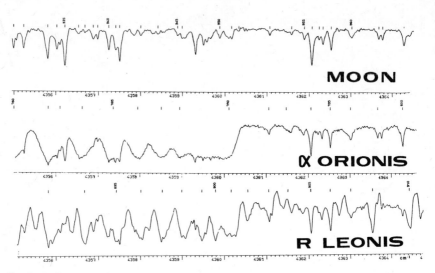

FIGURE 1. Spectrum of the Moon, α Orionis, and R Leonis with resolution of about 0.02 cm^{-1}, taken by Connes and Michel.[1]

the lines of the stellar atmospheres are resolved, and further resolution is not needed. It is true that there are quieter places in the universe than stellar atmospheres, such as inter-stellar clouds, where the linewidths are narrower than what one sees in these stars or the earth's atmosphere, but such techniques, resolving as small a spectral difference as about one two-hundredth of a wave number, are adequate to resolve even those lines. Connes entire spectrum of a star cannot be easily shown because, as he points out, to show the amount of the detail which is in the spectrum would require a length of paper about 200 metres long. The information is tabulated, available to specialists, and can justify an enormous amount of study.

For microwaves, at the other end of the spectral region to be discussed, we have parametric amplifiers and maser amplifiers which reach within an order of magnitude of the ultimate sensitivity, close enough that even the small residual emission from the atmosphere or from the ground can be the limiting factor in sensitivity. The resolution can be almost arbitrarily high because one uses heterodyne methods for which a bandwidth of any fineness can be achieved, although a very long time may be needed for measurement. Figure 2 shows very narrow spectral lines of water from a naturally occurring water maser in the nebulosity W49, These spectra were taken by Sullivan[2] at the Naval Research Laboratory on several different occasions about a month apart. They occur at about 24 000 MHz and have widths as small as a fraction of a kilohertz in some cases. They change from month to month because this maser is produced in a relatively local, very intense, and very active source. The spectral structure seen is all due to the Doppler

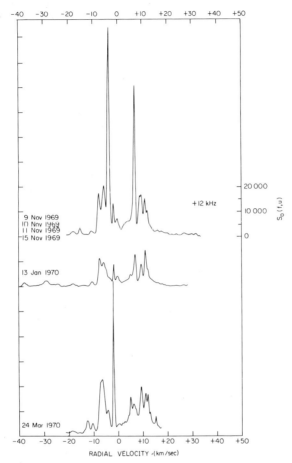

FIGURE 2. Emission from water masers in the astronomical object W49, showing variations in intensity with time (from Sullivan[2]).

effect, each line probably coming from a different region of a complex group of clouds. The intensity of activity suggests that stars are forming in this region, but its nature is not yet well understood.

Figure 3 shows a spectrum of ammonia in an interstellar cloud obtained by Mayer et al.[3] using a maser amplifier. Here the spectrum is rather weak; nevertheless the hyperfine structure of ammonia on either side of the principal line can be seen. These hyperfine component lines would, for an "optically thin" cloud, be still weaker than is observed by comparison with the central component. Their relative intensities are important in that they show that the cloud is in fact "optically thick" in spite of the fact that the total intensity of the lines is low. This indicates that the medium consists of a

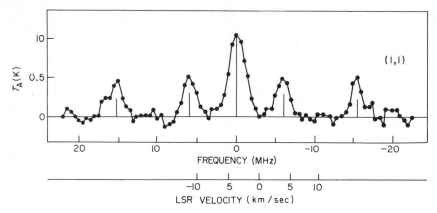

FIGURE 3. Spectrum of ammonia from a dark interstellar cloud showing hyperfine structure (from Mayer et al.[3]).

series of fairly dense blobs rather than a continuous cloud, which has been the more natural assumption.

For wavelengths somewhat shorter than the normal microwave region, i.e. the short millimetre or submillimetre range, techniques are not yet so well developed and are both insensitive and awkward.

Figure 4 shows a spectrum of ^{12}CO in Orion at 1.3 mm wavelength, where the atmosphere is still moderately transparent. This was made with heterodyne-type techniques involving frequency multiplication in a Schottky diode. The effective temperature of the line observed in this case is about 40 K—a

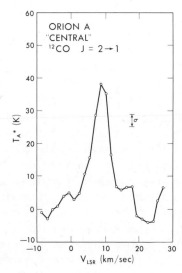

FIGURE 4. Spectrum of ^{12}C^{16}O from a dark cloud in Orion (from Goldsmith et al.[4]).

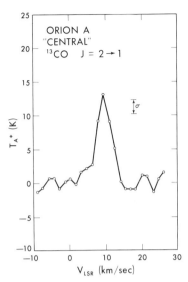

FIGURE 5. Spectrum of the rare isotopic species $^{13}C^{16}O$ from a dark cloud in Orion (from Goldsmith et al.[4]).

fairly warm, intense line. Figure 5 shows ^{13}CO in the same source. In spite of the relative rarity of ^{13}C, this line is also rather intense, which shows that this cloud, too, is optically dense. These spectra, obtained by Goldsmith et al.[4] are some of the earliest spectra obtained in this region. CO has already been studied extensively in the lower rotational transition, $J = 1 \rightarrow 0$, and there is obviously still more of its rotational spectrum as yet undetected to shorter wavelengths, which will be quite important in understanding the state of CO, very likely not a thermal equilibrium one, and the characteristics of such clouds.

Figure 6 shows a spectrum of the same rotational transition obtained by Phillips et al.[5] of the Bell Telephone Laboratories. This group has devised a clever heterodyne system using a solid of very fast thermal relaxation as a mixing detector. Such a system is not very easily tunable, so his measurement utilized a spatial sweep through Orion. This system has the low noise temperature of about 200 K. Its spatial sweep through the Orion cloud reveals a regular periodicity in the structure of the cloud, indicating that it is not a series of random blobs but periodic in structure. Recently this has been shown to represent a periodicity in velocity rather than density alone. There is nothing similar available at still shorter wavelengths; observations of shorter wavelength rotational transitions of CO from interstellar clouds do not yet exist.

Figure 7 shows one of the reasons why work at shorter wavelengths has been slow. Atmospheric transmission for visible radiation is good, and this is

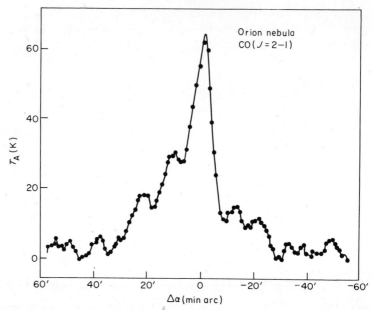

FIGURE 6. Variation in CO radiation intensity with position in the Orion nebula (from Phillips et al.[5]).

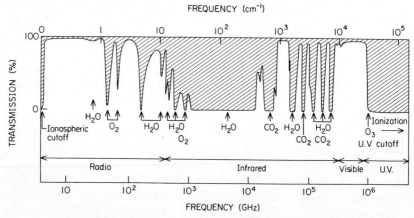

FIGURE 7. Absorption of infrared in the earth's atmosphere.

where most astronomy has been done. Near 10 microns there is a region which gives another good window, but at longer wavelengths transmission is poor. There is a partial window at 20 microns and a rather poor "window" near one third of a millimetre. Otherwise, the atmosphere is not cooperatively transparent. However, in spite of this difficulty there is obviously an enormous wealth of astronomical information to be somehow obtained within this span of three decades between a few microns and a few millimetres wavelength.

Almost everything we know about astronomy comes from spectroscopy, the measurement of electromagnetic waves—not quite all, but almost all. Astronomy is thus a type of applied spectroscopy. Furthermore, it has the peculiarity, as opposed to many interesting fields reported yesterday, that we cannot tamper with the object of study itself. We cannot send light through astronomical objects to obtain absorption spectra; we cannot boost the intensity of astronomical objects nor change their condition; we simply must observe what is there. This puts enormous demands on technique if we are to obtain all the information that is in fact present.

For the submillimetre region only a very limited amount of work can be done from the ground, primarily because of absorption by water vapour in the atmosphere, so that one necessity is to get above the atmosphere. This requires work from balloons, airplanes, or spacecraft. Figure 8 shows a small jet aircraft operated by NASA which has a 12-inch telescope looking out of its fuselage. A 12-inch objective is about as big as could be used without too serious structural troubles. The figure also shows Don Brandshaft of the University of California who is carrying out experiments in the

FIGURE 8. NASA plane equipped with 12-inch side-looking telescope for astronomy above the troposphere, D. Brandshaft with his equipment installed.

FIGURE 9. Interior of NASA plane for high altitude astronomy, showing detector dewar and side-looking telescope.

100 μm wavelength region. Figure 9 shows some of the interior, also including the small telescope and a cryogenic detector. This interior is crowded, but NASA has just put into service a larger plane, a C141, which is enormous by comparison and which carries a telescope of about one metre diameter. It will lend itself to a great deal of interesting work. The apparatus of Brandshaft and McLaren, utilizing a Fabry-Perot interferometer, has a maximum resolution of about 100—not high, but about as high as has so far been obtained at this wavelength. Obviously, a resolution of 100 is still quite inadequate for resolution of most lines. We badly need the type of resolution in this region afforded by heterodyne techniques. One can hope that detectors such as Schottky diodes, metal-oxide contacts, Josephson junctions, or photo conductive solids will ultimately allow good heterodyne detection. The local oscillators needed can clearly be constructed because there is a series of molecular lasers which can be used in this region.

At somewhat shorter wavelengths, in the 5, 10, 20 micron regions, substantially more work has been done. Even here, however, the first spectral lines outside of the solar system were detected only about five or six years ago. These regions are also not very highly developed, but work can be done from the ground. Infrared astronomical lines are not normally intense, and the weaker signals in these regions require long observation for good detection. In the microwave region astronomers observe for many days, even many weeks, to detect some of the weaker lines. In the infrared, generally such long times have not yet been attempted, but times of the order of hours are frequently used and longer observations are called for. This puts a considerable premium on higher sensitivity, but also on stability, reliability, and multiplexing techniques.

The Fourier-transform spectroscopy which Connes has used so successfully is quite attractive from the point of view of multiplexing; that is, one can observe many lines at the same time. It is particularly useful, however, only when noise comes from the detector and not from other sources. For example, if the noise comes from fundamental fluctuations in received photons, then a Fourier spectrometer has essentially the same speed as a very-narrow-band spectrometer with no multiplexing at all. This is simply because the wider band of the multiplex system increases the noise and it turns out that the total time required to observe a spectrum is exactly the same as if a single good detector is used to detect only a narrow band. In the 10 micron region, the noise usually has a still worse characteristic. For quite a while, astronomers have talked about *sky noise* at 10 microns because observations through telescopes characteristically shows systematic fluctuations which disturb any measurements attempted. Many of these variations actually have their origin in the telescope, for example in heat radiation from parts of the telescope. Most telescopes have simply not been designed properly for infrared work. In addition, there is some radiation from the sky—real *sky noise* and both sky and telescope give systematic variations which are much bigger than the fundamental "shot noise" due to photons. In such a case the multiplexing of a Fourier-transform system is a disadvantage, making such a system slower than one with a single detector and very narrow bandwidth.

To obtain a narrow pass-band of variable frequency for 10 μm radiation, we have used a tandem Fabry-Perot interferometer and filter wheel. Figure 10 shows a schematic diagram of such a spectroscope. The radiation traverses two Fabry-Perots which are swept in tandem and then a filter wheel before impinging on a photoconductive detector. Figure 11 shows the resulting pass band. The fine Fabry-Perot has many pass-bands within the region, the coarse Fabry-Perot has fewer, and the filter wheel has only one. The three in series result in a single narrow pass-band which can be swept in frequency by driving the Fabry-Perot spacing electronically. In this way one obtains the spectrum and avoids disturbance from some of the excess noise because only a very narrow band of radiation falls on the detector.

168 C. H. TOWNES

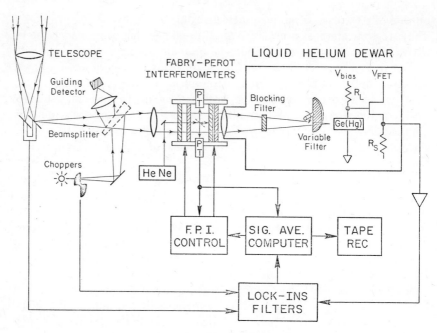

FIGURE 10. Schematic of tandem Fabry-Perot spectrometer (from Geballe[6]).

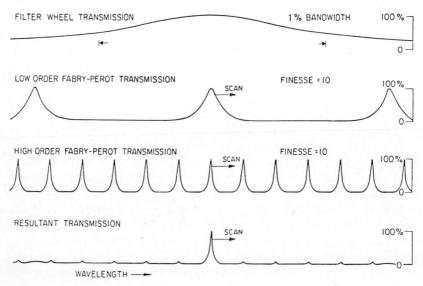

FIGURE 11. Pass-band of tandem Fabry-Perot spectrometer and its individual components:

The above method may seem awkward in that it cannot easily cover a wide spectral region. However, there is an enormous amount of information in the small range of frequency which can be covered. Table 1 illustrates why that is so. The lines of CO in the range of about five wave numbers at five microns are listed; about 34 lines can reasonably be expected to be found in a stellar atmosphere within this narrow range. These lines provide much information about excitation of various vibrational-rotation states, and about isotopic abundance of the several different isotopes of carbon and oxygen. Results with the above instrument are shown in Fig. 12, which illustrates spectra of several different stars using a resolution of about one

TABLE 1. CO Vibration-rotation lines in five cm^{-1} interval near 2142 cm^{-1}

Molecule	Transition	Rest Frequency	Line Number
C12 O17	5-4 R47	2144.70	
C13 O17	1-0 R23	2144.63	
C12 O16	7-6 R69	2144.55	1
C13 O16	5-4 R60	2144.52	2
C13 O18	3-2 R61	2144.49	
C12 O17	4-3 R34	2144.48	
C12 O16	5-4 R34	2144.34	3
C12 O18	5-4 R63	2144.21	
C13 O18	2-1 R45	2144.20	
C13 O16	1-0 R13	2144.04	4
C12 O18	2-1 R23	2144.94	5
C12 O16	6-5 R47	2143.85	6
C12 O18	3-2 R33	2143.81	
C12 O16	7-6 R68	2143.71	7
C13 O18	1-0 R33	2143.63	
C13 O16	4-3 R43	2143.45	8
C12 O18	4-3 R45	2143.30	
C13 O16	5-4 R59	2143.22	9
C13 O18	3-2 R60	2143.16	
C12 O18	1-0 R14	2143.07	10
C13 O16	3-2 R31	2143.03	11
C12 O18	5-4 R62	2143.03	
C13 O17	3-2 R44	2142.99	
C12 O16	4-3 R23	2142.95	12
C12 O17	5-4 R46	2142.86	
C12 O16	7-6 R67	2142.84	13
C12 O16	3-2 R14	2142.71	14
C12 O17	3-2 R23	2142.64	
C13 O17	2-1 R32	2142.48	
C12 O16	2-1 R6	2142.47	15
C13 O16	2-1 R21	2142.39	16
C13 O18	2-1 R44	2142.26	
C12 O17	4-3 R33	2142.09	
C12 O17	2-1 R14	2142.07	17
C12 O16	6-5 R46	2142.03	18
C12 O16	5-4 R33	2141.94	19
C12 O16	7-6 R66	2141.91	20
C13 O16	5-4 R58	2141.88	21

TABLE 1—continued

Molecule	Transition	Rest Frequency	Line Number
C13 O17	1-0 R22	2141.84	
C12 O18	5 4 R61	2141.81	
C13 O18	3-2 R59	2141.79	
C12 O17	1-0 R6	2141.58	22
C13 O16	4-3 R42	2141.43	23
C12 O18	3-2 R32	2141.39	
C12 O18	4-3 R44	2141.37	
C13 O18	1-0 R32	2141.23	
C12 O18	2-1 R22	2141.13	24
C13 O17	3-2 R43	2141.02	
C12 O17	5-4 R45	2140.98	
C12 O16	7-6 R65	2140.95	25
C13 O16	1-0 R12	2140.83	26
C12 O18	5-4 R60	2140.54	
C13 O16	3-2 R30	2140.54	27
C13 O16	5-4 R57	2140.49	28
C13 O18	3-2 R58	2140.40	
C13 O18	2-1 R43	2140.28	
C12 O16	6-5 R45	2140.16	29
C12 O16	4-3 R22	2140.09	30
C13 O17	2-1 R31	2140.04	
C12 O16	7-6 R64	2139.93	31
C12 O18	1-0 R13	2139.91	32
C12 O17	3-2 R22	2139.81	
C12 O17	4-3 R32	2139.67	
C12 O16	5-4 R32	2139.51	
C13 O16	2-1 R20	2139.50	
C12 O16	3-2 R13	2139.48	
C12 O16	1-0 P1	2139.42	

tenth of a wave number in the small range of about five wave numbers. The numbered lines correspond to those listed in the last column of Table 1. Different stars immediately show interesting differences. For example, on the right side of Fig. 12 one sees that the intensity of lines due to a highly excited vibration state is quite high for a hot star (the uppermost spectrum) and steadily decreases as one goes successively to the lower spectra in this figure of successively cooler stars. Isotopic species can also be seen in these spectra, but they are better illustrated by Fig. 13. The spectrum in Fig. 13 is of IRC + 10216, an infrared source in the form of a star with several shells of gas and dust around it. This figure shows the same region of the spectrum as does Fig. 12; the lower temperature produces many fewer lines, but some very important ones—for example, those of $C^{17}O$ and $C^{18}O$. On earth, O^{18} has about five times the abundance of O^{17}. It is clear that the relative abundance of ^{17}O to ^{18}O has been very much increased in this star, by at least a factor of 25 over the terrestrial value. Recent studies of interstellar material indicate that the earth's isotopic abundances, as arbitrary as they might seem

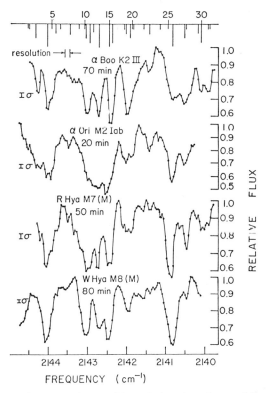

FIGURE 12. CO spectra from several stars with resolution about 0.1 cm^{-1} (from Geballe[6]).

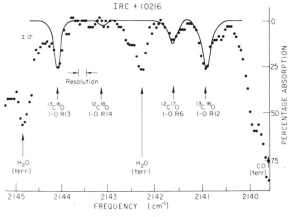

FIGURE 13. CO spectrum from the infrared star IRC + 10216 (from Rank et al.[7]).

to us, are in fact quite widespread throughout our galaxy. They are characteristic of most of the interstellar material in our galaxy, so that the peculiar abundances of ^{17}O are due to nuclear reactions within this particular star. It indicates, for one thing, that the star is a rather old one, and in its last stages of evolution. It also indicates at least one of the specific nuclear reaction processes which must be important. Doppler velocities of the expanding puffs of gas shells show that they have been emitted only during the last few thousand years.

Another type of 10 μm spectra of great interest is due to fine structure transitions of important ionic species. In this case, infrared observations allow us to measure characteristics of bright nebulosities, even though they may be obscured, in visible light, by dust clouds. The centre of the galaxy, for example, is surrounded by dust clouds and cannot be seen in visible radiation. The centres of galaxies are puzzling and important regions, and if we are to find out much about our own galactic centre, wavelengths larger than those in the visible region must be used. Infrared rays either from molecules or from the fine structure of atomic and ionic species can traverse the surrounding dust clouds and give us substantial information.

Figure 14 shows spectra of three planetary nebulae—bright nebulae excited by a star—taken by Holtz et al.[8] and showing a line of SIV. These represent the first high resolution spectral lines in the 10 μm region from

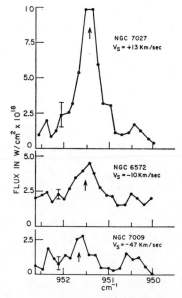

FIGURE 14. SIV fine structure radiation near 10 μm wavelength from three planetary nebulae, NGC 7027, NGC 6572, and NGC 7009 (from Holtz et al.[8]).

outside the solar system. Quite recently Aitken et al.[9] of University College, London, have obtained the fine structure of the line of NeII from the galactic centre and from it they obtained some parameters of the galactic centre. While such work is still in its very early stages, it should in the future be quite important. However, additional sensitivity and resolution is needed to detect and resolve such lines well, in order to measure details of Doppler velocities and to detect many of the less abundant ionic species. Success in this should yield much information on the state of excitation and conditions in the galactic centre. It is striking that these fine structure lines have never been detected in the laboratory, but their frequencies are predicted with some accuracy from ultraviolet measurements.

Until now, we have been discussing the spectral aspects of high resolution. In astronomy, high spatial resolution is also critically important. For spatial resolution, two interesting new techniques are being tried which have grown out of quantum electronics. Professor Harris discussed earlier the process of up-conversion in gaseous systems. We have been working with up-conversion of a somewhat older type, using nonlinear crystals to convert infrared waves to visible waves. In principle, up-conversion allows one to convert one quantum of infrared radiation into one quantum of visible light through non-linearities, and with the introduction of no new noise. In this way one can convert an entire infrared field of view into visible light, which can then be recorded as a photograph. Until now, essentially no photographs or no real pictures of infrared astronomical fields have been obtained, although astronomical infrared fields have been scanned to produce spatial representation of intensity. While up-conversion is a very powerful technique in principle, in fact it is exceedingly difficult and inconvenient to apply. One cannot up-convert a large range of frequencies nor a large field of view at any one time. The conversion is, furthermore, not highly efficient unless intense pulsed lasers are used to increase non-linearities. But such pulsed systems do not allow a very high duty cycle, and hence they can up-convert a field of view for only a small fraction of the time. There are thus many difficulties, but ones which may perhaps be overcome in the long run. Professor Harris' gas system is a very interesting one and may prove superior to crystal systems. However, for the moment it is not clearly superior and useful astronomy can already be done by up-conversion in crystals, and it is this which we have developed.

Figure 15 shows schematically a system which Boyd[10] of the University of California has put together. A krypton laser providing about 1/4 watt cw at visible wavelengths traverses the crystal, and is mixed with the infrared image from a telescope at ten microns. The laser light is then filtered out, leaving only the up-converted light which strikes an image converter and thus provides an image intense enough to be photographed. This apparatus should be able to take pictures of the atmosphere of Venus in a narrow

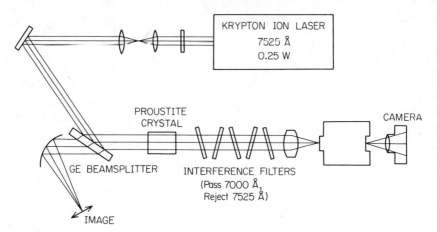

FIGURE 15. Schematic of infrared up conversion imaging system (from Boyd[10]).

spectral region of choice, so that one may look at the excitation of molecules in the relatively warm atmosphere of Venus with as much spatial resolution as a telescope can achieve through our atmosphere. 10 μm pictures of Mars or of the Moon can be still more easily obtained since their surfaces are warmer than the cloud tops of the Cytherean atmosphere.

While it may seem disadvantageous not to be able to up-convert all frequencies at one time, this property may be used to advantage in producing a spectrometer. Figure 16 shows resolution of HCl in the laboratory by an up-conversion system developed by Smith[11] of the University of California. Phase matching occurs in a limited wavelength range, a range which can be changed by changing the temperature or the orientation of the crystal. In Smith's apparatus the temperature of the up-converting crystal is swept slowly to scan the well-known spectrum of HCl. Lines of the two isotopic species of chlorine are seen well resolved, indicating a resolution of about 1 cm^{-1}.

The next high-resolution infrared system to be discussed returns to the field of high spectral resolution—much higher spectral resolution than the Fabry-Perot spectrometer described above can reasonably give. It uses heterodyne detection, which is so very favourable and universally used in the radio region. Heterodyne detection is not so favourable in the infrared region, and is only just starting to be applied in astronomy. In addition to the unavailability of local oscillators before lasers came along, heterodyne detection is less attractive at infrared wavelengths than it is in the microwave region because there is a basic noise in any kind of amplification or detection

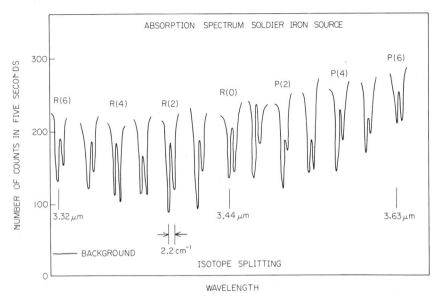

FIGURE 16. Infrared spectrum of HCl from up conversion spectrometer-detector (from Smith[11]).

which preserves phase—the basic noise associated with the uncertainty principle. This noise means that the noise temperature of the detector system must be comparable with or larger than $h\nu/k$, or about 1500 degrees for 10 μm wavelength. Thus, such a phase-preserving detector, looking out into space to detect some infrared radiation, must look through a noise background of about 1500 K or greater—a large amount of noise by comparison with that from wide-band detectors ordinarily used. However, in terms of sensitivity to power, a heterodyne detector can still have a high performance because a noise temperature represents a certain noise per unit bandwidth, and the bandwidth of a heterodyne detector can be made almost arbitrarily small. Thus, in terms of minimum detectable noise power, heterodyne detection can be more sensitive than any other present detection system in the infrared. Generally, we face an intermediate situation in which moderate bandwidths are needed and for which heterodyne detection is reasonably sensitive but not as sensitive as other available forms of infrared detection. Heterodyne detection of course becomes more favourable as we go to longer wavelengths. At five to 10 microns it may be somewhat marginal, but still useful. Whenever very high resolution is needed, heterodyne techniques should be considered, and even at 5 μm there are interesting astronomical sources which are intense enough to make heterodyne detection practical and rewarding. For wavelengths as long as 100 microns heterodyne detection is very much favoured, and of course in the microwave region it is usually the only sensible type of detection to use. Peterson et al.[12] of the University of California have used heterodyne detection at 10 microns, with a resolution of 3×10^7. This exceptionally high resolution could easily be made higher by the simple use of still narrower filter bandwidths. However, the amount of power available to be detected continually decreases with the bandwidth, and as a practical matter Peterson et al. used bandwidths of about 10 MHz which were adequately narrow to resolve well the narrowest line widths occurring in astronomical objects. Their system used a telescope of modest size, 30 inches in diameter, which focused an image on a copper-doped germanium photoconductive detector. A CO_2 laser local oscillator beam also falling on the detector mixed with the incoming signal. The germanium detector used had a bandwidth of about 1200 MHz, so that a bandwidth of twice 1200 MHz around the CO_2 line could be detected. This is very small compared with the total range of frequencies one would like to observe, but adequate to examine the contours of some particular lines under high resolution.

Figure 17 shows CO_2 lines in the atmosphere of Mars obtained by the above system, and never before resolved. These are due to $^{13}CO_2$, $^{12}CO_2$ being so saturated that its lines are much wider. The figure shows three different vibration-rotational lines in hot bands of CO_2 in the Martian atmosphere. Their widths are due about half to Doppler effects and about half to pressure broadening. The relative intensities of these lines and their shapes can give a substantial amount of information about the temperature and pressure distribution in the atmosphere of Mars. This preliminary measurement indicates that either the ^{13}C abundance or the atmospheric temperature was not as expected, and before long the work will be repeated for a more critical measurement and analysis.

Finally, I return to the question of ultimate spatial resolution in the infrared. In astronomy, it is clearly important to measure positions very accurately and to obtain very high angular resolution. Telescopes themselves do remarkably well, but atmospheric seeing does not characteristically allow an angle less than about one second of arc to be resolved. This resolution corresponds to the diffraction limit of a 10 cm aperture at visible wavelengths, which is as large a telescope as is usually needed for high visual resolution alone. Michelson improved on this resolution by using two small apertures separated by distances up to 20 feet and mounted on a single telescope—his famous stellar interferometer. The more recent interferometric technique of Hanbury-Brown and Twiss has also been used in the optical region. There has not, before this, been a high spatial resolution interferometer in the infrared region. Within the last decade, there has been much important work in the microwave region with what is known as long-baseline interferometry, whereby microwaves are received by two separate antennas, the two received

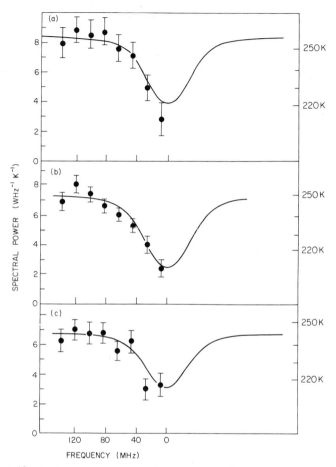

FIGURE 17. $^{13}CO_2$ absorption lines near 10 μm wavelength Mars, resolved by heterodyne spectroscopy (from Peterson et al.[12]).

signals recorded, and then compared. Perhaps the longest baseline used for such a system was from the western part of the United States to the Crimea. Thus, base lines comparable with the dimensions of the earth are being used, which are about as long as one can achieve without going out into space. In the microwave case, the two antennas are not generally tied together directly in phase; the radiation is compared with a very accurate clock in both positions, recorded on tape, and then the two tapes brought together to produce an interference pattern. This very powerful technique has given resolutions for radio sources even greater than that achieved by Michelson.

There are many sources which are primarily infrared—that is, they radiate relatively little microwave or visible radiation. Furthermore, even those sources which produce substantial visible and microwave radiation may have quite different distributions and sizes in the infrared region. For example, it is believed that much of infrared radiation from stars characteristically comes from a shell of material around the star, substantially larger than the star itself. Neither the actual size nor the shape of such infrared sources have been directly measured. There are other objects which we understand still more poorly, ones which are intense in the infrared region, but unknown in character. This makes high angular resolution in the infrared quite important for astronomy. I shall now describe a system by which we seek to achieve the needed resolution.

First of all, consider the size of telescope appropriate in the infrared to replace one of Michelson's single 10 centimeter apertures in his stellar spatial interferometer. The diameter of the telescope usable should increase, at least to a first approximation, as the wavelength, so at 10 microns one can use about 20 times as great a diameter as in the visible region, or about two metres. A standard theory based on the quite artificial assumption of random turbulence in the atmosphere shows that the usable size should increase as $\lambda^{6/5}$, which gives a diameter a factor of two still larger. Thus one can use a rather large single aperture and still obtain complete coherence over its surface. Similarly, the separation between the two apertures on the basis of this rather idealized theory, may increase as $\lambda^{6/5}$, so instead of 20 feet the aperture separation might be increased to about 800 feet, achieving about twice the angular resolution obtained by Michelson. In the microwave region it has been found that usable separations are much greater than is predicted by a similar argument, and the resolution obtainable much higher. It is presently not clear how much higher resolution might be obtained in the infrared region; this remains to be determined by some of the first experiments. In any case, one should expect to be able to use telescopes with dimensions of the order of metres and separated by a few hundred metres.

A 10 µm spatial interferometer, designed by Johnson et al.[13] is shown schematically in Fig. 18. Two separate telescopes each focus infrared radiation on a heterodyne detector, the local oscillator being a CO_2 laser. The signal obtained after this mixing is in the I.F. region and has a bandwidth of about 2400 MHz, about as large a bandwidth as can be easily achieved with present detectors. One laser serves as a local oscillator for one detector, and another one, kept in phase with the first but offset in frequency by 5 MHz, is local oscillator for the second. The interference pattern in the sky is therefore sweeping through the sky at 5 MHz—that is, a single lobe of the two-antenna pattern goes past a given star at a frequency of 5 MHz. This frequency is far above that of any atmospheric fluctuation, and is convenient for various technical reasons, but not very important in principle. The I.F. signals are brought together, added or multiplied, processed and averaged, so that

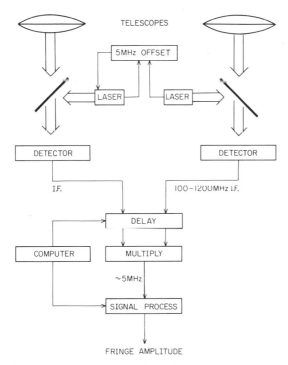

FIGURE 18. Schematic of 10 μm spatial interferometer (from Johnson et al.[13]).

interferences are obtained near 5 MHz.

Obviously, some intensity is lost by accepting only radiation in a frequency bandwidth as narrow as 2400 MHz. On the other hand, many problems are also eliminated. One problem with any interferometer of the Michelson type is to obtain a "white light fringe." For this, the precision with which path lengths through the two apertures must be matched is given by the velocity of light divided by about 10 times the bandwidth. In the optical region, the precision is thus about a wavelength of light. Such precision is difficult, particularly if one attempts to maintain large separations. For our bandwidth of about 2400 MHz, the precision needs only to be about a centimetre, which is reasonably easy to achieve by switching in various electrical delay lines in the R.F. region. Thus, the paths for two telescopes which are separated by a large distance may be adequately equalized before the signal is multiplied and processed to obtain an interference. An alternative interferometer could use a larger bandwidth and in principle obtain greater sensitivity—either by mounting two apertures on a single telescope or by using a very precise optical delay line. Alternatively a narrow bandwidth could be used with an optical delay line in order to reduce the precision required of the latter. How-

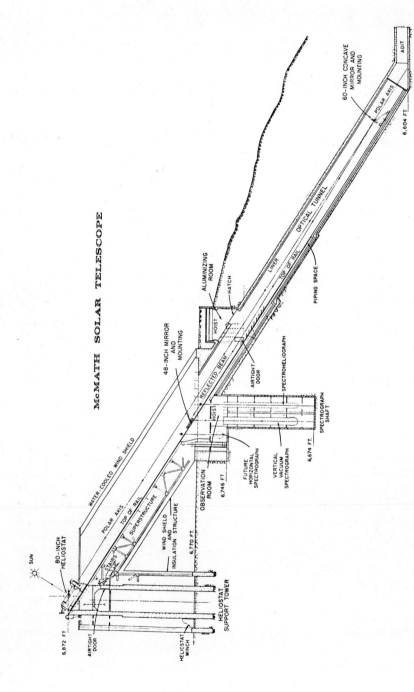

FIGURE 19. Diagram of the McMath Solar Telescope, Kitt Peak National Observatory.

ever, for large aperture separations, the heterodyne system appears cheaper and more flexible.

Figure 19 is a diagram of the McMath Solar Telescope. The actual telescopes used were two auxiliary 30-inch telescopes on either side of the main McMath telescope. Each has a 30-inch aperture coelostat, and the two are separated by about 18 feet. Their received light goes down a long tunnel, is reflected back again, then reflected again and into the observing room. The total optical path is somewhat longer than 100 metres in each of two telescopes. The seeing for such a telescope is somewhat worse than in most because of the long light paths, but these two telescopes have in other ways been very convenient, and we are very grateful for their being made available by the Kitt Peak National Observatory.

Figure 20 is a picture of the two auxiliary coelostats at the top of the telescope. The figure also shows the main telescope, of 60 inch aperture, along with the two smaller mirrors separated by 18 feet at the top of the long tube.

Figure 21 is a photograph of Michael Johnson and Albert Betz at the observing positions in the observing room. Here one sees the two helium dewars for maintaining proper temperature of the germanium detectors, the guiding optics, and arrangements for mixing in local oscillator beams.

FIGURE 20. Coelostats of the McMath Solar Telescope, including two auxiliary coelostats used for the 10 μm interferometer.

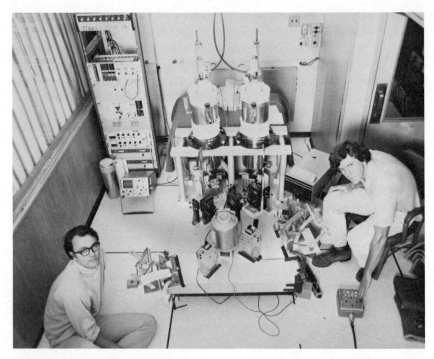

FIGURE 21. Michael Johnson and Albert Betz at observing positions of the 10 μm interferometer, showing cryogenic detector systems.

Figure 22 represents a signal from Mercury, which is a suitable test source for the system because it produces rather intense 10 μm radiation and is easy to guide on. Mercury is inconveniently large, since it subtends about six seconds of arc, whereas our interferometer resolves a small fraction of a second of arc. This gave a visibility of the fringes which was much less than unity but still provided a large enough signal for an adequate test. As Mercury rises above the horizon, the rate at which interference occurs gradually changes. While interference is occurring at 5 MHz because of the 5 MHz laser offset, there is a natural rate of fringe change due to the rotation of the Earth (or the rising of Mercury) and associated with change of effective baseline length. As Mercury rises, this provides a fringe rate in the range of a few to a few tens of Hertz. Figure 22 represents a record of the Fourier spectrum of the frequencies in the total signal. The signal has been recorded on magnetic tape for later processing so that there is a permanent record of all of the spectrum of interest. It is then analysed at leisure with a computer. The signal is clearly not of a completely defined frequency because of both detector noise and fluctuations in the atmosphere. The effect of fluctuations in path length through the atmosphere is one of the principal phenomena we

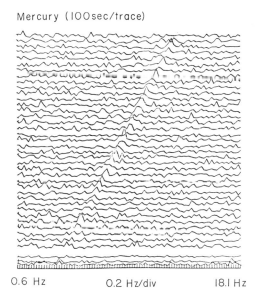

FIGURE 22. Fourier spectra of interference signal from the planet Mercury, showing variation in frequency of signal with time (from Johnson et al.[13]).

wish to investigate in order to examine the future potential of such interferometry. Each trace in Fig. 22 represents about 100 seconds of averaging, and the frequency and fringe fluctuations both differ for each period.

Figure 23 shows the total signal from a Fourier analysis of the entire trace with shifts in frequency made so that the theoretical fringe frequency expected would be superimposed throughout the observing period for 4000 sec. The coherence of the observed signal is remarkable. The Fourier analysis has resolution of 1/4000 Hertz. Much of the signal falls in a single frequency channel of this width. It will be seen that there is also some signal at nearby frequencies. Additional analysis shows that some phase fluctuations occur in times as short as one or two seconds, although there is a long-term coherence of phase over the entire 4000 seconds of observation. For several reasons, it may in fact be even better than this figure indicates. This remarkable coherence found is very encouraging for the future of any type of infrared spatial interferometry. It has much to do with the obtainable signal-to-noise ratios, with how well one can determine the absolute position of an object, and with the possibility of complex aperture synthesis.

The signal strength agrees with theoretical expectations when the measured fluctuations are taken into account, and is adequately strong to allow measurement of the brighter infrared stars. However, it needs further improvement. For this we need both larger telescopes and better detectors. The detectors

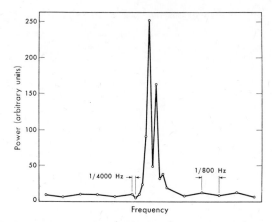

FIGURE 23. High resolution Fourier spectrum of the interference signal of Mercury analysed in a way which keys the expected signal frequency at the same position throughout 4000 seconds of observation (from Johnson et al.[13]).

used were home-made and had a sensitivity about 25 times the fundamental limit previously mentioned, corresponding to a noise temperature of $h\nu/k$. Detectors are known which have sensitivities not more than about four times that limit and we hope to obtain some of this quality. With detectors of the best presently known quality and 60-inch aperture telescopes, the signal-to-noise ratio would be improved by a factor of about 25. How long a base-line can practically be used is still not clear. The technique lends itself to distances as large as hundreds of kilometres. The most useful work will probably be done with telescope separations about one kilometre. The telescopes need only be joined, as mentioned above, by an R.F. cable in order to obtain the interference signal. How quickly the fluctuations will increase with distance is not yet known. How well one can determine the position of a star, or the rate of rotation of the Earth as it sweeps past the star every day, will depend on just how narrow the ultimate line width is and whether there is a central more-or-less fixed relative phase between the two telescopes, as there probably will be at least for the shorter distances.

While the present test tells us much about the quality of the atmosphere, it is not a very complete test because the telescopes themselves introduce phase fluctuations between the two paths. In fact, tests on fluctuations in the telescope show that they are of approximately the same magnitude as was observed on Mercury. To obtain atmospheric fluctuations alone, those within the telescopes will eventually be removed by using a monitoring laser beam. The atmospheric fluctuations may possibly then appear still smaller and more favourable.

REFERENCES

1. Connes, P., and Michel, G. (1974), *Astrophys. J.*, **190**, L29.
2. Sullivan, W. T. III, (1973), *Astrophys. J. Supp.*, **25**, 393.
3. Mayer, C. H., Waak, J. A., Dheung, A. C., and Chui, M. F. (1973), *Astrophys. J.*, **182**, L65.
4. Goldsmith, P. F., Plambeck, R. L. and Chiao, R. Y. (1975), *Astrophys. J.*, **196**, L39.
5. Phillips, T. G., Jefferts, K. B., Wannier, P. G., and Ade, P. A. R. (1974), *Astrophys. J.*, **191**, L31.
6. Geballe, T. (1974), Thesis, University of California at Berkeley.
7. Rank, D. M., Geballe, T. R., and Wollman, E. R. (1974), *Astrophys. J.*, **187**, L111.
8. Holtz, J. Z., Geballe, T. R., and Rank, D. M. (1971), *Astrophys. J.*, **164**, L29.
9. Aitken, D. K., Jones, B. and Penman, J. M. (1974), *Mon. Not. R. Astr. Soc.*, **169**, 35.
10. Boyd, R., private communication.
11. Smith, H., private communication.
12. Peterson, D. W., Johnson, M. A. and Betz, A. L. (1974), *Nature*, **250**, 128.
13. Johnson, M. A., Betz, A. L., and Townes, C. H. (1974), *Phys. Rev. Lett.*, **33**, 1617.

Chapter 10

Photoelectron Spectroscopy in the Study of Molecular Orbitals

by W. C. Price, King's College, London WC2R 2LS, England

INTRODUCTION

High-resolution ultraviolet photoelectron spectroscopy has proved to be the most revealing technique yet developed for the elucidation of the electronic structure of molecules. It has demonstrated the existence and properties of molecular orbitals in a way which has not been possible hitherto by conventional photon spectroscopy. Thus it has given a new depth to our understanding of the nature of the chemical bond. Techniques have been developed for obtaining the spectra of both ionic and covalent molecules and transient as well as stable molecular species. Progress has also been made in the interpretation of particular band features in terms of molecular properties of the ground or the various excited ionized states. These advances will be illustrated in what follows by reference to a systematic study of the simplest groups of molecules in which their orbital structures are developed from those of related united atoms from which they can theoretically be derived by subdivision of the positive charge centres.

For the photo-ejection of an electron from an orbital in a molecule we have the equation

$$I_0 + E_{vib} + E_{rot} = h\nu - \tfrac{1}{2}mv^2$$

where I_0 is the "adiabatic" ionization energy (i.e. the pure electronic energy change) and E_{vib} and E_{rot} are the changes in vibrational and rotational energy which accompany the photoionization. The energy ($h\nu$) of the incident photon is known from the wavelength of the radiation used which is usually resonance radiation from a discharge in an appropriate inert gas. The 58.4 nm resonance line of helium which has an energy of 21.21 eV is commonly used. The range of kinetic energies ($\frac{1}{2}mv^2$) of the photoelectrons ejected from a molecule depends on the pattern of vibrational states accessible within the energy of the incident photon and their intensity distribution depends on the probabilities of the transitions to these vibrational states. The calibration of the photoelectron spectrum is achieved by simultaneously recording the spectrum of a gas of spectroscopically known ionization potential. This avoids the effect of contact potentials etc. in the electron spectrometer.

Figure 1a shows the photoelectron spectra of H_2 and D_2 obtained by irradiation with the He I (58.4 nm) line. Figure 1b gives the potential curves of H_2 and H_2^+. When an electron is removed from the neutral molecule the nuclei find themselves suddenly in the potential field appropriate to the H_2^+ ion but still separated by the distance characteristic of the neutral molecule. The most probable change is thus a transition on the potential energy diagram from the internuclear separation of the ground state to a point on the potential energy curve of the ion vertically above this. This is the Franck-Condon principle and it determines to which vibrational level of the ion the most probable transition (strongest band) occurs. The energy corresponding to this change is called the vertical ionization potential I_{vert}.

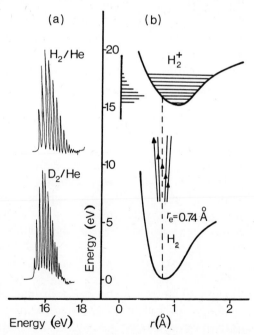

FIGURE 1. (a) Photoelectron spectra of H_2 and D_2; (b) potential energy curves showing spectra plotted along ordinate.

Transitions to vibrational levels on either side of I_{vert} are weaker, the one of lowest energy corresponding to the vibrationless state of the ion corresponding to I_{adiab}. The photoelectron spectra of H_2 and D_2 are given in Fig. 1a and that of H_2 is also plotted along the ordinate of Fig. 1b. It will be appreciated that its band envelope reflects to a considerable extent the displacement of the potential curve on ionization.

It is possible to calculate the change in internuclear distance on ionization from the intensity distribution of the bands in the photoelectron spectrum. Clearly when a bonding electron is removed, the part of the photoelectron spectrum corresponding to this will show wide vibrational structure with a frequency separation that is reduced from that of the ground state vibration. The removal of relatively nonbonding electrons, on the other hand, will give rise to photoelectron spectra rather similar to those of the monatomic gases and little if any vibrational structure will accompany the main electronic band. The type of vibration associated with the pattern obtained when a bonding electron is removed can usually be identified as either a bending or a stretching mode and this can throw light on the function of the electron in the structure of the molecule, that is, either as angle forming or distance determining. From the band pattern it is frequently possible to calculate values of the changes in angle as well as changes in internuclear distance on ionization, and so the geometry of the ionic states can be found if that of the neutral molecule is known. Changes in bond lengths or angles can be calculated with about 10% accuracy for photoelectron band systems which show structure by using the semi-classical formula

$$(\Delta r)^2 = \ell^2 (\Delta \theta)^2 = 0.543 \left[I_{vert} - I_{adiab} \right] \mu^{-1} \omega^{-2}$$

where r and ℓ are in Å, θ in radians, μ in atomic units, and ω the mean progression spacing in 1000 cm^{-1}.

PHOTOELECTRON SPECTRA OF HYDRIDES ISOELECTRONIC WITH THE INERT GASES

As examples of the features mentioned above and also to illustrate how the orbitals of isoelectronic systems are related to one another we shall now discuss the photoelectron spectra of those simple hydrides for which the united atoms are the inert gases Ne, Ar, Kr and Xe—that is, the atoms obtained by condensing all the nuclei into one central positive charge. A molecule of hydrogen can of course be considered as being formed from an atom of helium by the subdivision of its positive charge into two equal units. These are held together by the electrons in a $(1s\sigma)^2$ orbital of somewhat ellipsoidal distribution which is derived from the spherically symmetrical $(1s)^2$ He orbital and is centred in between the two protons. Similarly hydrides such as HF, H_2O, NH_3 and CH_4 can be thought of as being formed from Ne by successively partitioning off protons from its nucleus. Because of the nature of orbitals and the fractional charge division the changes are not quite so simple as in the case of hydrogen. The solutions of the wave equation—the wave functions or orbitals—arising from the progressive subdivision of the positive charge pass continuously from the atomic to the molecular cases with a gradual splitting of the p^6 orbital degeneracy as the molecular fields are set up. The effect of splitting off a fraction of the central positive charge on the $2s^2$ orbitals will be to distort them slightly towards the extracted protons; i.e. causing "bumps" to arise in the bond directions. The effect on the 2p orbitals will be somewhat different. Schematic diagrams of the orbital changes are indicated in Fig. 2. The photoelectron spectra shown in Fig. 3a, b, c and d reveal this orbital subdivision in a striking way. Figure 3a shows the spectra of the halogen acids which is the first stage in this process. The spectra of the corresponding inert gas (i.e. the united atom) is inserted on the records for comparison. The triply degenerate p^6 shell is split into a doubly degenerate π^4 shell and a singly degenerate σ^2 shell. The π^4 shell is non-bonding and represented by two bands in the photoelectron spectrum which are of equal intensity and show little vibrational structure. The σ shell is formed from the p orbital along which the proton is extracted and gives a negative cloud which binds the proton to the residual positive charge. This $p\sigma$ orbital therefore gives rise in the photoelectron spectrum to a simple progression of bands with a separation corresponding to the vibration frequency of the $^2\Sigma^+$ state of the ion except where the structure is lost by predissociation.

The second stage of partition leads to the molecules H_2O, H_2S, H_2Se and H_2Te. If two protons were removed colinearly from the united atom nucleus the linear triatomic molecule so formed would not have maximum stability, since only the two electrons in the p orbital lying along the line will then be effective in shielding the protons from the repulsion of the core. By moving off this line, the protons can acquire additional shelter from the central charge and from themselves through shielding by one lobe of a perpendicular p orbital (see Fig. 2c). The electrons in this orbital then become "angle determining" as distinct from the two previously mentioned p electrons which are mainly effective in determining the bond separations of the hydrogen atoms

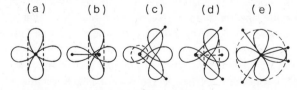

FIGURE 2. Schematic diagrams of 2p-orbital structures of (a) Ne, (b) HF, (c) H_2O, (d) NH_3 and (e) CH_4.

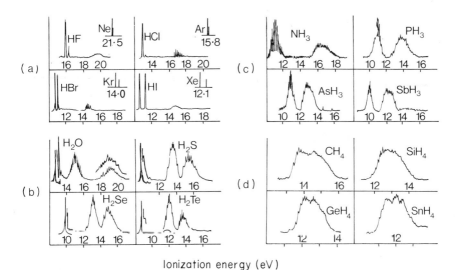

FIGURE 3. Photoelectron spectra of the hydrides isoelectronic with Ne, Ar, Kr and Xe obtained with He(21.22 eV) radiation.

from the central atom. The remaining p orbital, because of its perpendicular orientation to the plane of the bent H$_2$X molecule, can affect neither the bonding nor the angle, that is, its electrons are nonbonding. These expectations are borne out by the photoelectron spectra shown in Fig. 3b. These spectra show how in all H$_2$X molecules the triple degeneracy of the p^6 shell of the united atom is completely split into three mutually perpendicular orbitals of different ionization energies. The lowest ionization band is sharp with little vibrational structure. The second has a wide vibrational pattern which turns out to be the bending vibration of the molecular ion. The third also has a wide, vibrational pattern with band separations greater than those of the second band. These separations can be identified as the symmetrical bond stretching vibrations of the ion. The geometries of these three ionized states can be calculated from the band envelopes and pattern spacings. Only small changes are associated with the band of lowest i.p. Large changes of angle accompany ionization in the second band, which in the case of H$_2$O causes the equilibrium configuration of this ionized state to be linear. In H$_2$S, H$_2$Se and H$_2$Te the spectra show that the angles are 129°, 126° and 124° respectively. The third ionized state is one in which the internuclear distances are increased but little change occurs in the bond angle. The integrated intensities of all three bands are roughly equal indicating that each originates from the ionization of a single orbital.

The partitioning of three protons from the nucleus of an inert gas molecule leads to a pyramidal XH$_3$ molecule. One p orbital is directed along the axis

FIGURE 4. Qualitative model showing the $1t_2$ and $2a_1$ orbitals of CH_4 with reference to the enveloping cube. Protons are small spheres with black equatorial bands, the p_x, p_y, p_z components of t_2 are represented by pairs of spheres with longitudinal bands and the 2a orbital is represented by the darker central area.
Note the relative sizes of the orbitals have been distorted to bring out the geometrical features (see text).

of symmetry and provides the shielding which causes the molecule to have a nonplanar geometry, that is, it is angle determining. The other two p orbitals are degenerate and provide an annular cloud of negative charge passing through the three XH bonds and thus mainly determine the bond distances. The structure pattern on the first band of the photoelectron spectrum (Fig. 3c) can be assigned to bending vibrations and indicates that the molecule flies to a symmetrical planar configuration without much change in the bond distances when an electron is ionized from this orbital. The second band in the photoelectron spectrum shows Jahn-Teller splitting which is consistent with its being doubly degenerate. Although it has limited structure, such structure as can be observed corresponds to changes in bond stretching without much change in bond angle.

Finally the photoelectron spectra of the XH_4-type molecules show that the p^6 shell of the united atom has changed to another triply degenerate shell with orbitals of t_2 symmetry which can be roughly described as being formed from p_x, p_y, p_z orbitals oriented parallel to the edges of the enveloping cube. This is indicated in the qualitative model shown in Fig. 4. In this model the protons are represented by small spheres with black equatorial bands, the p_x, p_y, p_z orbitals are represented by 3 larger pairs of spheres with longitudinal bands and the a_1 (tetrahedrally distorted 2sC) is the central more darkly shaded volume for which in fact a small van der Waals' space-filling model of methane was used. In practice the s and the p orbitals interpenetrate considerably and to obtain a more realistic approximation to the electron density distribution in the molecule the spheres representing the "p" lobes should be pushed in radially until they touch at the centre. The radial distances of the protons should also be considerably reduced. Since these adjustments would make it difficult to show the geometrical properties of the orbitals the more open arrangement has been retained in the model but it should be viewed in the light of the above provisos.

The contours of the $(t_2)^{-1}$ photoelectron bands show the presence of Jahn-Teller splitting, and the structure on the low-energy side shows that on ionization the molecule moves by contraction along one side of the enveloping cube towards a square coplanar configuration. In the heavier molecules, for example, SnH_4, the large spin-orbit splitting of the heavy atom influences the structure. The spectra also show that the movement toward coplanarity

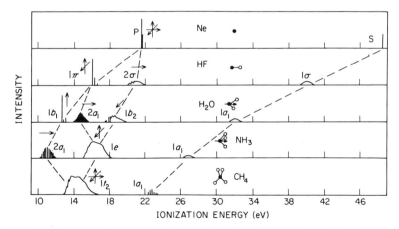

FIGURE 5. Diagrammatic spectra of the hydrides formed by proton withdrawal from neon.

becomes progressively pronounced. Further details on the spectra of these hydride molecules are given by Potts and Price.[1, 2]

It has been mentioned that in the process of partitioning off protons from the neon nucleus to form the second row hydrides, the 2s orbitals are distorted though to a lesser extent than the 2p orbitals. They are deformed from spherical symmetry in the direction of the extracted proton. They thus have "bumps" in the bond directions and in this way provide an "s" contribution to the bond. This is evident in the vibrational structure of the photoelectron bands associated with these orbitals, for example the $(a_1)^{-1}$ band systems are simple progressions in the totally symmetric XH stretching vibrations reduced in frequency by about 25% from their values in the neutral molecules. Figure 5 shows how the orbitals of the hydrides of F, N, O and C are formed by proton withdrawal from neon.

PHOTOELECTRON SPECTRA OF THE HALOGEN DERIVATIVES OF METHANE

It is convenient to discuss at this point the photoelectron spectra of some "single bond" molecules in which some of the atoms have additional groups of nonbonding electrons. The bromomethanes, whose photoelectron spectra are illustrated in Fig. 6, are good examples of this molecular type. It can be seen that some bands in their spectra are relatively sharp and therefore can be associated with the nonbonding electrons, while others are broad and clearly arise from electrons in strongly bonding orbitals. The orbital assignments of the latter bands are indicated schematically in the figure. A compari-

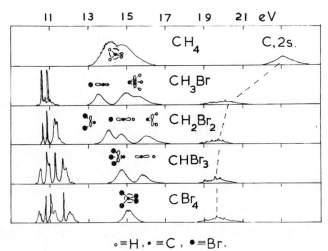

FIGURE 6. Photoelectron spectra of the bromomethanes.

son of these bands with those of the hydrides of the same symmetry, e.g. CH_3Br and NH_3, CH_2Br_2 and H_2O etc., shows that the orbitals around the carbon atom are split up in a very similar way as in the hydrides by the departure from tetrahedral symmetry brought about by the halogen substitution. This gives some support to the old concept that around an atom in a stable molecule there should be a closed (inert gas) shell of electrons. It shows further how the degeneracy of the p^6 group of these electrons is split by the fields arising from the different substituents, the splitting of the degeneracy being complete in the case of the methylene halide as it is for water in the hydrides. The sharp bands in the region of 11 eV can be readily associated with orbitals containing $4p\pi$ Br electrons which are split by spin-orbit interaction. In the case of methyl bromide two sharp bands (with very weak accompanying vibrational structure) are split by the magnitude of spin-orbit coupling constant which is 0.32 eV for bromine. Further splitting occurs as the number of bromine atoms increases and eight bands can be seen in CBr_4 corresponding to the eight nonbonding "$p\pi$" orbitals present in this molecule. A detailed account of the analysis of the photoelectron spectra of the halides of elements in groups III, IV, V and VI of the periodic table has been given by Potts et al.[3]

SPECTRA OF IONIC MOLECULES

The binding in the molecules we have so far considered is of the covalent type. In such binding the valence electrons are approximately equally attracted to either of the fragments into which the united atom has been partitioned. When this is not so as for example in the case of LiF for which the united atom is magnesium (configuration $1s^2\ 2s^2\ 2p^6\ 3s^2$) partitioning off a 3+ fragment will remove first the weakly bound $3s^2$ electrons to form the $1s^2$ shell of Li^+ while the more tightly bound 2p electrons will tend to remain in the $2p^6$ shell thus forming an F^- fragment. The field tending to pull over the third electron is clearly dependent on the difference between the ionization potential of the metal atom $I(M)$ and the electron affinity of the halogen atom $E_a(X)$. Values of these differences for the alkali halides range between 2.33 eV for LiF to 0.28 eV for CsCl whereas for covalent bonds they are much larger, e.g. H_2 or CH_3-Cl where the differences would be about 13 and 6 eV respectively. The $I(M) - E_a(X)$ differences affect the bonding through the changes which arise in the short-range repulsive forces associated with the overlap of the closed shells of the ions. A study of the electrons in these closed shells by photoelectron spectroscopy might therefore be expected to give direct information concerning the repulsive factors which largely determine the strength of this type of binding.

Photoelectron spectra of the chlorides, bromides and iodides of Na, K,

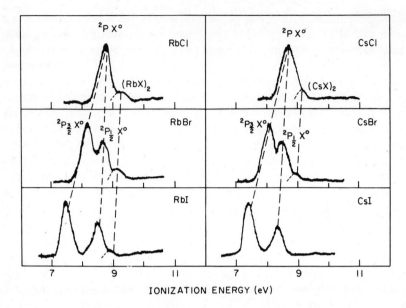

FIGURE 7. HeI spectra of caesium and rubidium halides.

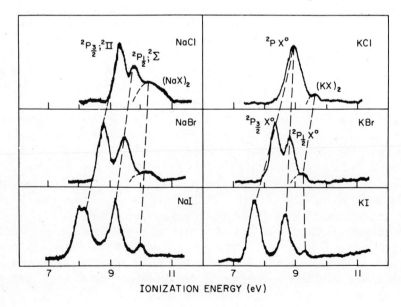

FIGURE 8. HeI spectra of sodium and potassium halides.

Rb and Cs recorded in the vapour state by a molecular beam technique are shown in Figs 7 and 8.[4] All the spectra correspond to the removal of an electron from the X^- $(p)^6$ shell of the molecule. This produces an M^+X^0 ion in which the ionic bond has been destroyed leaving only a small residual bonding due to the attraction between the M^+ ion and the induced dipole in the X^0 atom. The photoelectron spectra of the M^+X^0 molecules correspond to transitions to the various states of the M^+X^0 ions the relative separations of which may be deduced by consideration of the polarization of the X^0 atom by the M^+ ion i.e. by Stark splitting of the ^2P state of the X^0 atom in fields of different strength for each particular halide. On the basis of Mulliken's treatment[5] we can represent the probable splitting of the X^0 ^2P state by the energy level diagram shown in Fig. 9. The states of the X^0 ^2P atom and their relative energies should correspond to those observed for the M^+X^0 ion in the photoelectron spectra of M^+X^-. Describing the states of M^+X^0 in terms of the states of the X^0 atom, for the "zero field" situation, the spectra should show structure corresponding to $^2P_{3/2}$ and $^2P_{1/2}$ atomic X^0 states, the separation of the bands being just $3/2 \times$ the spin-orbit coupling constant ζ for the ^2P state of the free halogen atom. The spectra obtained for the Cs and Rb halides do in fact correspond to just this situation as might be expected from their low values of $I(M) - E_a(X)$ which are 0.280 and 0.563 eV for CsCl and RbCl respectively i.e. if any negative change is placed between M^+ and X^0 the attraction towards M^+ is largely counterbalanced by that due to the electron affinity of X^0. The separation found for the $^2P_{3/2}$ and $^2P_{1/2}$ bands is as predicted although in the case of the chlorides the two bands expected are not resolved. For the potassium halides (Fig. 8) the

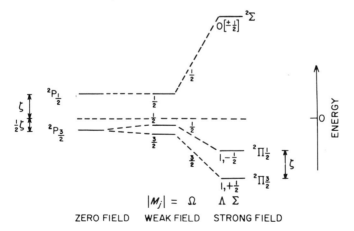

FIGURE 9. Correlation of the various energies of ionized states formed by Stark splitting of the X^0, ^2P state.

spectra are beginning to show evidence of a polarizing field and show greater splittings than are found for free atoms. The spectra of the sodium halides provide examples of polarization ranging from the weak field to the strong field situation. For NaI the $^2P_{3/2}$ state is split into $M_j = 3/2$ and $1/2$ components and the $^2P_{3/2}-^2P_{1/2}$ splitting is slightly greater than $3/2\,\xi$. In the spectrum of NaBr the $^2P_{3/2}-^2P_{1/2}$ splitting is increased to 0.65 eV as compared with a value of 0.46 eV for $3/2\,\xi$, while for NaCl the splitting is increased from 0.11 eV to 0.46 eV. These indications of the presence of large fields agree with the relatively large $I(M) - E_a(X)$ differences which are 1.525, 1.775 and 2.075 eV respectively for NaCl, NaBr and NaI.

Features to the high ionization energy side of the $(p)^{-1}$ bands observed in a number of the spectra are attributed to ionization of the $(MX)_2$ dimer which will be present in increasing amounts for the lighter halides. Comparison with the reliable values for the polymer (solid crystal) which are now available[6] shows that the energies required to ionize the dimers and polymers are greater than those needed for the monomer. The greatest increment, occurs in going from monomer to dimer the subsequent changes being an order of magnitude smaller. For example the first i.p. of NaCl is 9.34 eV, that of $(NaCl)_2$ is 0.96 eV higher while that of $(NaCl)_n$ crystal is raised still further by 0.15 eV. These increments are clearly due to increases in the polarization stabilization of the ionized states of Cl^- arising from the increasing population of its environment by polarized ions. At present we are still at an early stage in the study of ionic binding by photoelectron spectroscopy, but valuable new information is already being accumulated.
[7, 8, 9]

PHOTOELECTRON SPECTRA OF "MULTIPLE" BONDED DIATOMIC MOLECULES

To discuss "multiple" bonded systems in which molecular orbitals are formed by p atomic orbitals combining in the "broadside on" as well as in the "end on" configuration we shall take as examples the molecules N_2, NO and F_2. Their photoelectron spectra are shown in Fig. 10. The orbitals upon which their electronic structures are built are the in-phase and out-of-phase combination of the appropriate 2s and 2p orbitals. These are given schematically in the insert. For N_2 the electronic configuration is $(\sigma_g 2s)^2 (\sigma_u 2s)^2 (\pi_u 2p)^4 (\sigma_g 2p)^2$, the orbitals being in order of decreasing ionization energy. With the exception of $(\sigma_u 2s)$, all of these might be expected to provide excess negative charge density between the two nuclei and thus to account for the strength of the N_2 "triple" bond. The additional electron of NO has to go into a $\sigma^* 2p$ orbital which is largely outside the nuclei and therefore antibonding, and so in a loose analogy N_2 is the inert gas of diatomic molecules

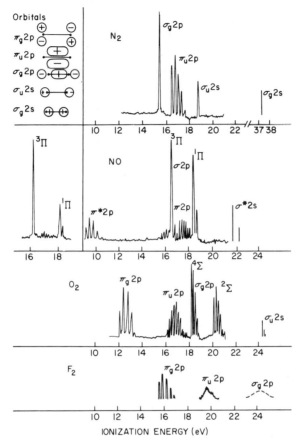

FIGURE 10. Photoelectron spectra of N_2, NO, O_2 and F_2.

and NO the corresponding alkali metal, since it contains one electron outside a closed shell of bonding electrons. An inspection of the photoelectron spectrum of N_2 in which the ionized states corresponding to removal of the different orbital electrons are marked, shows that by far the most vibration accompanies the removal of the π_u electron. Thus at the internuclear distance of neutral N_2, the nuclei are held mainly by the negative cloud of the $(\pi_u 2p)^4$ electrons. The short-distance bonding character of these π electrons results in the nuclei being pulled through the $(\sigma 2p)^2$ cloud so that this σ orbital is as much outside the nuclei as between them and therefore supplies no bonding at the N–N equilibrium separation. It can be understood readily from the geometry of their overlap that the bonding of p electrons in the broadside on (π) arrangement optimizes at shorter internuclear distances than that in the end on (σ) position. In N_2 the nuclei are in fact on the inside of the minimum of the partial potential energy curve associated with the $(\sigma 2p)^2$ electrons. This orbital is thus in compression, and its electrons have both their bonding and their binding (ionization) energies reduced. In NO, O_2 and F_2 the presence of the additional electrons in antibonding $\pi 2p$ orbitals causes the internuclear distances to be relatively longer than they are in N_2, and it can be seen from Fig. 10 that the $\sigma 2p$ bands of these molecules have progressively more vibrational structure as the internuclear distance approaches more closely that separation for which the bonding of the $(\sigma 2p)^2$ orbital is optimized. The associated σ bands move through the $\pi 2p$ systems to higher ionization energies in accord with their increased effective bonding power.

The features discussed above can be illustrated by considering orbital potential energy curves as illustrated in Fig. 11. In the orbital approximation of molecular electronic structure, the complete dissociation energy curve of a molecule can be split up into the partial orbital bonding curves which give the orbital bonding and the ionization energies over the whole range of internuclear distances. Different types of orbital have their potential minima at different internuclear distances. These do not coincide with the actual equilibrium internuclear distance r_e of the molecule which is determined by the minimum of the sum of the orbital energies taken at each value of r. At any particular r the electrons in different orbitals are at different relative positions in their orbital binding energy curves. These orbital curves are not of course directly determinable but have been drawn to be, as far as possible, consistent with the experimental facts. For instance the $(\sigma_g 2p)^2$ binding curve has been drawn as a near Morse curve to have an r_e of about 1.4 Å a dissociation energy of 4 eV and an $\omega_e = 1200 \text{ cm}^{-1}$. The curve for $(\pi_u 2p)^4$ can then be found for $r > r_e$ by subtracting the σ_g curve from the observed dissociation curve of N_2 on the assumption that at large distances the

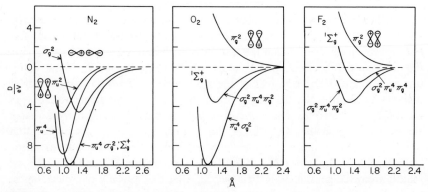

FIGURE 11. Orbital contributions to the potential energy curves of the $^1\Sigma_g^+$ states of N_2, O_2, and F_2.

$(\sigma_g 2s)^2 (\sigma_u 2s)^2$ orbitals do not contribute to total bonding. For the present purpose it has been assumed that this is also true at smaller values of r. The $(\pi_u 2p)^2$ curve can then be obtained by halving the $(\pi_u 2p)^4$ curve obtained by the above subtraction. The $(\pi_g 2p)^2$ repulsive curve can be obtained by plotting the difference between O_2, $X^1\Sigma_g^+$ and N_2, $X^1\Sigma_g^+$ which should check with the $(\pi_u 2p)^2$ curve of which it is a reflection at large r. The antibonding power of the antibonding orbital is only slightly greater than the bonding power of the bonding orbital (i.e. the former is proportional to $S(1 - S)^{-1}$ and the latter to $S(1 + S)^{-1}$, where S is the overlap integral). Similarly the $(\pi_g 2p)^2$ curve of fluorine can be obtained by plotting the difference between F_2, $X^1\Sigma_g^+$ and O_2, $X^1\Sigma_g^+$ or alternatively by halving the differences between the N_2, $X^1\Sigma_g^+$ and the F_2, $X^1\Sigma_g^+$ curves.

It will be noted that the optimum bonding energies of $(\sigma_g 2p)^2$ and $(\pi_u 2p)^2$ are roughly equal. This might be thought to contradict the fact that the single, double and triple bond energies of ethane, ethylene, and acetylene are not in the ratio of 1:2:3, but in the ratio 1:1.76:2.41. However it is readily appreciated that the figures indicate that the σ bond must lose about 60% of its bonding by compression to smaller internuclear distances when forming part of the triple bond. Because of the steep slope of the repulsive part of its curve, this loss would rapidly increase with further reduction in internuclear distance. Thus in N_2 the $(\sigma_g 2p)$ orbital has lost nearly all its bonding power at the equilibrium internuclear distance and its photoelectron band has the features characteristic of ionization from a nonbonding orbital. The rapid increase in its bonding power with increasing r is evident from the increasing vibrational structure of the $(\sigma 2p)^{-1}$ systems in NO and O_2 (see Fig. 10). On the other hand, the $\pi_u 2p$ electrons in N_2 find themselves at internuclear distances only slightly larger than those of their potential minimum and are thus strongly bonding. As the equilibrium internuclear distance increases in passing from N_2 to NO, O_2 and F_2 as electrons are added in antibonding repulsive orbitals, the $\sigma_g 2p$ orbital acquires bonding ultimately becoming the basic single bond in F_2. The value of the dissociation energy of F_2 (1.6 eV) is less than the optimum $(\sigma_g 2p)^2$ bond energy of about 4 eV, because, as already mentioned, the bonding power of the $(\pi_u 2p)^4$ orbitals is more than offset by the antibonding power of the $(\pi_g 2p)^4$ orbitals which are filled in this molecule.

Other interesting points illustrated by Fig. 10 are that the spacing in the $\pi 2p$ antibonding bands in NO and O_2 (first systems) are larger than those of the $\pi 2p$ bonding bands (second systems). This is to be expected since the removal of an antibonding electron increases the vibration frequency while that of a bonding electron decreases this frequency. Another interesting feature is that the separation of the first and second bands, which reflects the overlap between the out-of-phase and in-phase $(\pi 2p)$ orbital, rapidly decreases with increase in internuclear distance. For NO, O_2 and F_2 these separations are 7.5, 4.5 and 3.0 eV respectively. This indicates the reduction in "multiple" (π) bonding as the effect of the increasing number of π antibonding electrons reduces and annuls the bonding of π bonding electrons by increasing the interatomic distance and reducing the orbital overlap.

The insert of the 16 to 18 eV region of NO in Fig. 10 shows the $^3\Pi$ and $^1\Pi$ bands with an intensity ratio of 3:1, and thus illustrates how closely the intensities follow the statistical weights of the ionized states, agreeing with the number of channels of escape open to the electrons. Similar remarks apply to the $^4\Sigma$ and $^2\Sigma$ bands of O_2, the integrated intensities of which are in the ratio of 2:1. A further interesting feature which these multiplets illustrate is the greater bonding of the states of lower multiplicity. Since the lower multiplicity corresponds to the states of antisymmetric spin function, it is associated with the symmetric (summed) space coordinate wave function of the orbitals between which the spin interaction is occurring. The additional overlap to which this gives rise results in greater bonding relative to states of higher multiplicity where the total space coordinate wave function is obtained by subtracting those of the interacting states.

Space restriction prevents the extension of this discussion to triatomic and polyatomic molecules for which the reader is referred to a review article by Price.[10] In conclusion it should be mentioned that this paper has been written in terms of the orbital approximation. All that it is permissible to consider as representing an electronic state of an atom or a molecule is a certain spatial probability of electron density. This can only be acted upon with dipole radiation to excite or eject one electron if the charge density is divided into orbitals which have the symmetry properties of the system. Thus while these particular orbitals are most suitable for discussing spectroscopic transitions they may not be the ones most suitable for considering chemical changes. For a fuller discussion of this aspect of the electronic structure of atomic and molecular systems the reader is referred to an article by Daudel.[11]

REFERENCES

1. Potts, A. W., and Price, W. C. (1972), *Proc. R. Soc.*, **A326**, 165.
2. Potts, A. W., and Price, W. C. (1972), *Proc. R. Soc.*, **A326**, 181.
3. Potts, A. W., Lempka, H. J., Streets, D. G., and Price, W. C. (1970), *Phil. Trans. R. Soc.*, **A268**, 59.
4. Potts, A. W., Williams, T. A., and Price W. C. (1974), *Proc. R. Soc.*, **A341**, 147.
5. Mulliken, R. S. (1930), *Rev. Mod. Phys.*, **2**, 60.
6. Poole, R. T., Liesegang, J., Jenkin, J. G. and Leckey, R. C. G. (1973), *Chem. Phys. Lett.* **23**, 194.
7. Berkowitz, J. (1972), *J. Chem. Phys.*, **56**, 2766.
8. Berkowitz, J., and Dehmer, J. L. (1972), *J. Chem. Phys.*
9. Berkowitz, J., Dehmer, J. L., and Walker, T. E. H. (1973), *J. Chem. Phys.* **59**, 3645.

10. Price, W. C. (1974), *in* "Advances in Atomic and Molecular Physics," Bates, D. R. ed., pp. 133-171, Academic Press, New York.
11. Daudel, R. (1973) *in* "Wave Mechanics," Price, W. C., Chissick, S. S., and Ravensdale, T. eds, pp. 66-71, Butterworth, London.

General References

Eland, J. H. D. (1974), "Photoelectron Spectroscopy", Butterworth, London.

Siegbahn, K., Nordling, C., Fahlman, A., Norberg, R., Hamrin, K., Hedman, J., Johanson, G., Bergmark, T., Karlson, S. E., Lindgren, I., and Lindberg, B. (1967), "ESCA - Atomic, Molecular and Solid State Structure studied by means of Electron Spectroscopy"., North Holland Publ., Amsterdam.

Siegbahn, K., Nordling, C., Johansson, G., Hedman, J., Heden, P. F., Hamrin, K., Gelius, U., Bergmark, T., Werme, L. O., Manne, R., and Baer, Y. (1969), "ESCA—Applied to Free Molecules"., North Holland Publ., Amsterdam.

Turner, D. W., Baker, A. D., Baker, C. and Brundle, C. R. (1970), "Molecular Photoelectron Spectroscopy" Wiley (Interscience), New York.

Chapter 11

Spectroscopy Within the Cell*

by Britton Chance, P. Leslie Dutton, John S. Leigh, Johnson Research Foundation, University of Pennsylvania School of Medicine, Philadelphia, Pa., 19174, U.S.A.

INTRODUCTION

In our studies, we have used a variety of techniques which are problem-oriented rather than method-oriented and have often chosen the technique at hand which works and perhaps not the one of ultimate capability. Thus, I shall mention one or two new methods, but more often applications of older ones.

I should like first to review the problem of the "spectroscopist inside the cell" very briefly. One of the problems faced by the spectroscopist working on biological problems is the very low concentration of chemicals: 10^{-6}M, is an average concentration, but this may be as low as 10^{-9}M. Secondly, there is an inherent instability of life processes. We are dealing with a variety of steady states, each of which may be transitory and soon shift to a new steady state; indeed, oscillations in the concentration of intracellular chemical species may occur. Thirdly, a narrow temperature range is available between 35°C and 40°C for physiological studies, although many studies of intracellular constituents are carried out at room temperature (23°C) and may range much lower in the presence of non-aqueous "anti-freezes"—but this is a new technique in itself. Fourthly, an intrinsic complexity of the system arises from a wide range of overlapping chromophores. Fifthly, there is a truly remarkable range of speed of biochemical reactions, varying from diffusion-controlled reactions to the light-activated primary reactions of photosynthesis in the picosecond time range. There is, finally, a complexity of chemical pathways, many of which remain to be elucidated.

*This work was supported by USPHS GM-12202 and NINDS-10939.

One approach, which has been used by famous workers in the field such as Otto Warburg,[1] Otto Meyerhoff,[2] and Sir Hans Krebs,[3] is to extract, purify and then reassemble the biological material. This works well and offers an insight on a variety of chemical pathways which are appropriate to the intact system; however, this approach leaves great gaps in our knowledge as to the relative importance of the pathways and whether or not they really function in the cell as they do in the test tube.

There is a minority who have attempted the more difficult task of leaving the system intact and untampered. As Dr. Townes has indicated, this is often impossible—the system is too far away or too small—and one must observe directly by sensitive spectroscopic methods similar processes in cells and tissues with a spatial resolution limited to about 1 micron by the wavelength of light. This allows small molecular aggregates such as intracellular organelles to be resolved with no details of molecular structure.

Historically, cell spectroscopy was first employed by MacMunn,[4] who used a prism spectroscope as an ocular attachment to a microscope. Keilin[5] observed with the greatest excitement the increases of intensity in the absorption bands of cell pigments, which he termed "cytochromes", as the oxygen available for cell respiration decreased. Indeed, the cytochromes, of which I shall speak at some length, are the electron transfer catalysts of cell respiration. It was Dr. Robert Hill who first observed cytochromes in photosynthetic systems by means of the microspectroscope[6a] and this is a method which he and Dr. E. F. Hartree use to great advantage to this day.[6b]

Photoelectric spectroscopy, especially employing methods that compensate for the highly scattering and absorbing nature of the living cells and tissues in relation to their small absorbancy changes, reveals as much detail as visual spectroscopy and has, of course, the advantage of greater speed and sensitivity. Various methods were developed in the early 1950's by Drs. L. Duysens in Utrecht,[7] H. Lundegårdh in Sweden[8] and ourselves.[9] Our method compensated for Rayleigh scattering by referring the spectroscopic changes to a nearby reference wavelength which was preferably an isosbestic point for the absorbancy change to be measured. This idea can in fact be traced to Tyndall and was used by Dr. G. Millikan in the 1930's.[10]

A most important step from the physiological point of view was the identification of the ATP-forming organelle that contains the cytochromes. In mammalian cells—more properly, eukaryotic cells, those that have their reproductive apparatus segregated in a nucleus—this organelle is termed the "mitochondrion". In Fig. 1, the mitochondria in a bat wing muscle are shown by electron microscopy; they look like grapes interspersed between the contractile apparatus. The economy of nature has provided up to 40% of the tissue weight just for the energy-conserving apparatus which forms ATP for this very rapidly contracting cricothyroid muscle.[11]

FIGURE 1. Electron micrograph of bat wing muscle (courtesy of D. Fawcett).

As the mitochondria became better identified, their biological function also emerged, in particular the oxidation of the tricarboxylic acids to water and carbon dioxide by the Krebs cycle, with the formation of six molecules of ATP, the "undevalued" currency of the energy system. This is a remarkable efficiency; the slower stages of anaerobic breakdown of glucose to pyruvate in the cytoplasm yield about three molecules of ATP, and the citric acid cycle, about thirty.

The totality of spectroscopically identifiable electron transfer components of the mitochondrial respiratory chain includes many more than Keilin's cytochromes. These components may be identified by fluorescence, absorption and electron paramagnetic resonance spectroscopy. Overlapping spectra may be resolved in time by kinetic methods, or in oxidation-reduction state by electrometric titrations. The latter were developed by Mansfield Clark[12] and have been applied in detail to complex biological systems by Dr. Robert Hill[13] and more recently by Drs. Dutton and Wilson in our laboratory.[14]

There are flavin enzymes, nucleotide enzymes, quinones, iron-sulphur proteins, iron-containing cytochromes, copper-containing proteins—fifteen to twenty respiratory carriers in all. A most interesting feature is that they fit into redox families of similar electron-carrying potentialities. While all such categorizations are ultimately confounded by detail, they serve to clarify the overall picture. Figure 2 shows three groups of carriers: one at the midpotential of NADH and the citric acid cycle intermediates such as malate, citrate and isocitrate, about -300 mV; a second, more positive group at the level of succinate, around 0 mV; and finally, a set at the level of cytochrome c, about $+200$ mV. The differences in midpotential between these groups are equivalent to about one molecule of ATP formed per two electrons transferred; in other words, energy can be conserved in the form of ATP by electron transfer across these potential gaps and, indeed, that is where ATP seems to be made in the respiratory chain. The general idea is that there is a redox ballast at three main points in the chain; two of these points, at the level of NADH and succinate, serve as physiological entry points for electrons, and perhaps protons, into the system. Thus, with NADH as the principal substrate for the passage of two electrons through the entire respiratory chain to oxygen, three ATP molecules are produced; from succinate, two ATP's; and if electrons are artificially fed into the chain at the cytochrome c level (via TMPD and ascorbic acid, for example) then the yield is only one ATP per two electrons.

Between the groups of carriers at about the same mid-potential, there are components of alternating mid-potential, for example, the iron-sulphur proteins at energy-conservation Site II, which can equilibrate in a system which conserves free energy and nevertheless permits the efficient formation

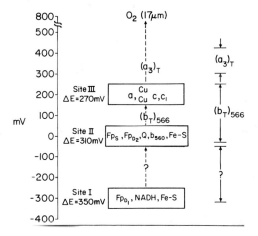

FIGURE 2. The electron carriers of the respiratory chain arranged according to their oxidation-reduction potentials.[14]

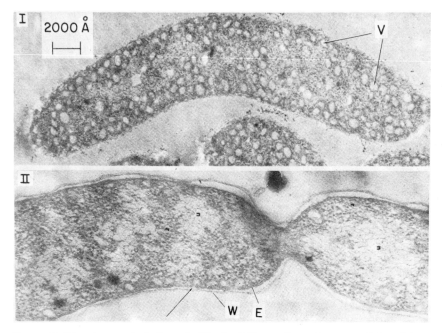

FIGURE 3. The photosynthetic bacteria Rhodospirillum rubrum (courtesy of Dr. Cohen-Bazire).

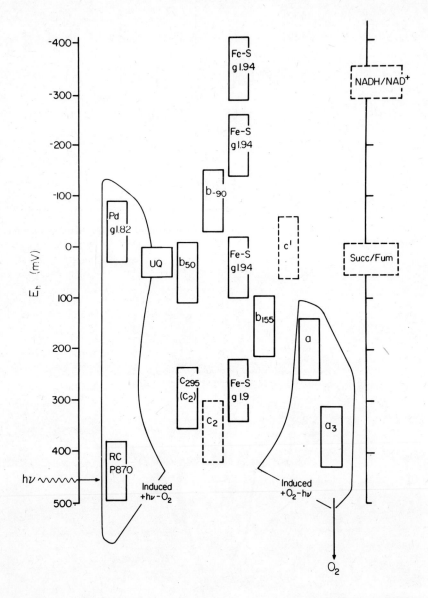

FIGURE 4. The redox potentials and organization of the respiratory carriers of the photosynthetic electron transfer chain.[14]

of ATP. This is perhaps a useful matter of principle. The way in which ADP is phosphorylated for ATP is not exactly known; it could be through a direct coupling of the respiratory components of alternating mid-potential to an ATP-synthesizing enzyme or it could be by way of ion gradients generated by the operation of such components. Both may be involved. In either case, thermodynamics cannot and has not been ignored (see Ref. 14). This contribution is focused upon the kind of electron transfer reactions that lead to the formulation of ATP in the mitochondria.

One way of approaching the system, over and above the oxidation-reduction states of the carriers, is through molecules that inhabit membranes and are responsive to localized or delocalized charge separation. Several workers have considered the accessory pigments of the photosynthetic system, such as the carotenoids, as such indicators—an approach that has been followed particularly by Witt and his group in Berlin[15] and by Jackson and Crofts in Bristol.[16] These carotenoids appear naturally in plants and photosynthetic bacteria, and we will consider them to be models for this system of charge indicators.

Figure 3 shows a micrograph of *Rhodospirillum rubrum*, a bacterium that lives mostly in mud and has achieved an infrared sensitivity by means of a chlorophyll that absorbs 800–1000 nm illumination very effectively, instead of the more usual chlorophyll of higher plants, activated at about 700 nm. Grown aerobically in the dark (bottom) the cells are featureless; grown in the light in the absence of air, they are full of small vesicles which are generated by involutions of the membrane. Breaking the cell leads to these involutions forming "chromatophores", which are the bacterial equivalent of small particles derived from mitochondria; they contain pigments that function in electron transfer and cellular respiration as do those of mitochondria. Simple grinding or sonication of the bacterial cells yields a suspension of chromatophores which are complete and functional, with a membrane 50–100 Å thick around a vesicle 600 Å in diameter.

Here again, as shown in Fig. 4, spectroscopy and redox potential titrations may be used to separate the classes of carriers, and the three mid-potential groups—NAD/NADH, succinate/fumarate and a third, more positive potential group—at cytochrome c_2, the photosynthetic analogue of cytochrome c are apparent. In the photosynthetic system, the activated reaction centre takes the place of oxygen and cytochrome oxidase as the electron acceptor and the associated ATP-producing site is omitted. This reaction centre produces both a positive and a highly negative potential by an electron—hole separation so that it can act as both ends of the electron transport chain—i.e., as both citric acid cycle and oxygen—for cyclic electron flow in the illuminated system. The bacteria are adapted to function from a modified citric acid cycle as well (including the glyoxylate cycle of

Kornberg[17]). We shall consider here their light-activated way of life and examine the elementary steps which occur upon activation by an 860 nm photon.

In photosynthetic bacteria such as these, the carotenoids may serve as potential indicators of localized or delocalized charge separations. They are conjugated molecules which undergo a shift of their absorption spectrum when the cell is illuminated. Figure 5 represents a light-minus-dark difference spectrum of the carotenoid effect, discovered by Duysens[7] and subsequently studied by Witt[15] in green plants and in bacteria by Smith and Ramirez[18] and Smith and Baltscheffsky,[19] and more recently by Jackson and Crofts.[16] This band shift is thought to be linear with respect to the diffusion potential induced by an ion gradient across the membrane; it seems that the carotenoid exhibits an electrochromic or Stark effect (see Ref. 22).

SOLVATO- AND ELECTROCHROMIC RESPONSES OF CAROTENOIDS AND MEROCYANINES

Figure 6 illustrates the equations for electrochromism which were described by Platt in a simplified form[20] and later by Bücher and his colleagues[21] in a more detailed equation, where the coefficients of the linear and squared terms of the electric field represent the permanent and induced dipole moments. Schmidt and Reich[22] have studied the carotenoid lutein in an artificial lipid bilayer across which they could apply a potential. They found that this carotenoid showed only the squared term; the permanent dipole response had cancelled out and so is of no interest in interpreting the data, in surprising contrast to the initial interpretation of the work of Witt[15] and of Crofts.[16]

KINETIC STUDIES OF CAROTENOID RESPONSES

The great speed of optical methods, when applied to the study of carotenoids and cytochromes, may serve to fill in what Martin Kamen[23] has termed "the spectrum of ignorance" in the photochemistry of biology, going from the very shortest times of 10^{-15} sec to very long times, on the order of 10^3 sec. As Kamen pointed out (cf Fig. 7), the microsecond region is fairly well known but the nanosecond region has been largely unexplored and the picosecond region inaccessible. Here we shall describe studies of the photosynthetic process from 5 psec onwards as a part of a joint report on work by my colleagues Dutton and Leigh at the Johnson Foundation, and Rentzepis, Netzel and Kaufmann at Bell Laboratories.[24, 25, 26] But first we shall look at the steady state picture.

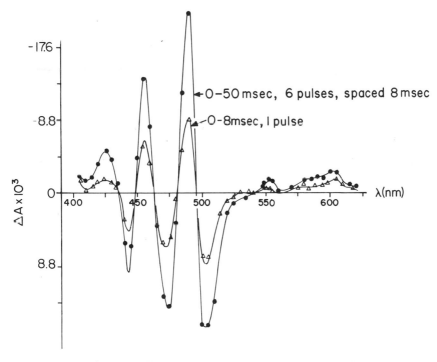

FIGURE 5. Light-minus-dark difference spectrum for the carotenoid response in *Rps. spheroides* (Dutton, P. L., unpublished data).

$$b_L \qquad\qquad b_R$$
$$O(=CH-CH=)_n CH-NH_2$$

$$O^-(-CH=CH-)_n CH=NH_2^+$$

$$\longleftarrow V_{L\ R} \longrightarrow$$

$$E^2 = E_I^2 + (b_L - b_R + V_{L\ R})^2$$

$$h\Delta\nu = \Delta\vec{\mu}\cdot\vec{F} - 1/2\,\Delta\alpha F^2$$

FIGURE 6. Simplified equations for electrochromic response according to Platt,[20] Bücher[21] and Cheng.[40]

Dutton's first sketch (Fig. 8) identifies the structural rearrangements that occur on illumination. Dr. Blasie in our laboratory has located the reaction centre in the membrane by X-ray diffraction in oriented multilayers, and the location shown here conforms to his data.[27] The reaction centre contains a chlorophyll dimer; the reaction centre can be isolated and broken apart into three subunits, two of which contain bacterial chlorophyll and a colourless unit, which may contain the electron acceptor, which is likely to be a quinone associated in some way with a non-heme iron;[28, 29] (For discussion, see Ref. 30).

In steady state illumination, the acceptor is reduced and shows a characteristic EPR spectrum at $g = 1.82$; the oxidized chlorophyll has an EPR spectrum of a free radical ($g = 2.0026$) and a bandwidth appropriately narrow for a dimer.[31] Cytochrome b is shown to be within the membrane and the two cytochrome c_2 hemes are located on the inside of this section of the 50–100 Å membrane. The quinone is considered to participate in electron transfer; it is a highly lipophilic, lipid-soluble substance which might be useful for ferrying electrons or hydrogen atoms between the primary acceptor or succinate and cytochrome b, within or across the membrane. Electrons can flow from the primary acceptor to the ubiquinone, which binds a proton; electron and proton are carried to cytochrome b, which interacts with cytochrome c_2 in an energy-linked reaction whose details are as yet unresolved. Cytochrome c_2 serves to reduce the membrane-associated, light-oxidized reaction centre chlorophyll, completing the cycle. Therefore, in the steady state the electron transfer pathway has an electron passing through this system and back through the reaction centre in a cyclic fashion so that the cytochromes are alternately oxidized and reduced.

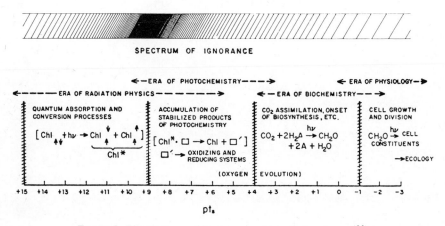

FIGURE 7. The spectrum of ignorance on biological processes.[23]

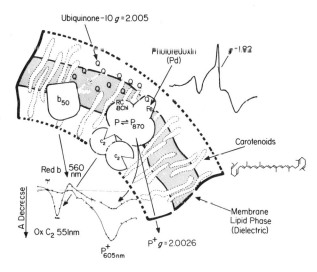

FIGURE 8. A schematic diagram of the arrangement of the components of the chromatophore membrane.[14] See Figs. 9–12 for explanations of symbols.

In order to explore the very early kinetic relationships of this cyclic system, we can use a spectrophotometer with mode-locked laser excitation to study the carotenoid band-shift 5 psec after excitation. I shall not discuss in detail the system of Rentzepis[24, 25, 26] which is similar to that already described by Professor Kaiser.[32] In brief, there is a Pockels cell which selects a single pulse of 530 nm excitation. A portion of this radiation is converted into a broad band from 500 to 540 nm by Stokes and anti-Stokes conversion. There is a two-channel echelon so that pathways for sample and reference wavelengths are available, one through the suspension and the other through an air path. The echelon gives radiation 5 psec before to 40 psec after the excitation pulse. The transmission along the two pathways is analysed by a 1 m focal length monochromator, and the absorption is registered on a Vidicon at the exit slit. Thus, the two spectra can be subtracted at all wavelengths to give the kinetic and spectral traces of Fig. 9.

We can examine the carotenoid band-shift and also the reaction centre triplet state which may be regarded as a charge transfer complex in the dimer and a precursor of the oxidized reaction centre dimer. This persists onto the microsecond time range. Figure 9A shows a typical trace of the formation of a carotenoid-associated band at 510 nm. The trace of Fig. 9B is thought to represent a charge transfer complex formed from the initial excited singlet state, acting as a precursor of the oxidized state. The decay of the excited charge transfer triplet state is interpreted as the course of the electron transfer from the bacterial chlorophyll to its acceptor; prior reduction of the primary acceptor prevents the decay of the precursor via the oxidation route and its lifetime lasts into the nanosecond range, where it decays through another form of the biradical which can be detected by EPR. The latter may be a ground state triplet or a biradical in which charge recombination occurs by tunnelling.

The rather large carotenoid-associated absorption increase with a maximum at 510 nm has been interpreted[26] as an electrochromic or Stark shift response to the initial charge separation across the chlorophylls of the dimer. This idea now seems less likely; the carotenoid shift may be mainly, if not entirely, due to a reaction of carotenoid with antenna chlorophyll in the chromatophore membrane, and the optical response is that of carotenoid triplet formation. Evidence for this comes from the observation that carotenoid associated with isolated reaction centres does not undergo any change in the picosecond time range; the carotenoid of the "PM8 mutant" of Sistrom and Clayton[33] which does not contain reaction centres, still shows a response, although somewhat less. Also, prior chemical oxidation of the reaction centre bacteriochlorophyll does not greatly alter the carotenoid response, although carotenoid transients are seen in tens of picoseconds under prolonged oxidative conditions. While more study is needed, Fig. 9C shows a possible scheme for these early events.

If, instead of picosecond excitation and readout, a Q-switched laser pulse is used, the primary photochemical electron transfer to the primary acceptor has already occurred during the excitation, as shown in Fig. 10. The carotenoid shift in this case requires a full separation in the reaction centre of oxidized chlorophyll, P^+, from reduced acceptor, X^-, as indicated by the EPR traces at $8°K$ which show the kinetics of reduction of the acceptor, followed at these temperatures by electron backflow, presumably by tunnelling, to restore the system to its initial state.

Figure 11 shows that the carotenoid band-shift can be further enhanced if a positive charge (hole) can be moved inwards towards the centre of the vesicle to cytochrome c, causing its oxidation. The electron is donated to the positive side of the reaction centre and there is a corresponding change in the carotenoid kinetics. This slow phase corresponds to the reduction of the oxidized reaction centre, and the oxidation of cytochrome c. The kinetic traces are similar. Diffusion will bring an additional cytochrome c molecule into action, to give another slow phase of electron transfer.

The next step in membrane energization, shown in Fig. 12, is electron transfer from the light-reduced, negative end of the primary acceptor system to the membrane quinone molecules, which are initially oxidized. At the same time that they accept an electron, a hydrogen ion is accepted. By putting

A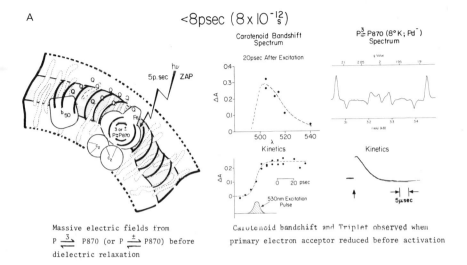

Massive electric fields from
$P \xrightleftharpoons{3} P870$ (or $P \xrightleftharpoons{\pm} P870$) before dielectric relaxation

Carotenoid bandshift and Triplet observed when primary electron acceptor reduced before activation

B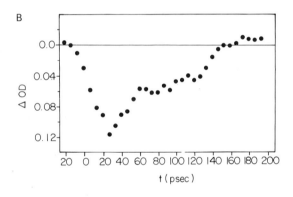

C The equation for initially reduced donor and oxidized acceptor is:

$$c_2^{2+}P\text{-}P \ (Fe\text{-}Q) \xrightarrow{h\nu} c_2^{2+}P^*\text{-}P \ (Fe\text{-}Q) \xrightarrow{\leq 7ps} c_2^{2+}P^{\pm}P^-(Fe\text{-}Q)$$

$$\xleftarrow{\sim 120\,ps} c_2^{2+} P^{+}\text{-}P \ (Fe \ \bar{Q}) \xrightarrow{30\mu sec} c_2^{3+}\text{-}P\text{-}P \ (Fe\text{-}Q)^-$$

Prior chemical reduction of FeQ prevents the 120 psec electron transfer and results in the visualization of the triplet or biradical character of the dimer by EPR.

FIGURE 9. A, B, C. Spectroscopic responses of the reaction centre as time-resolved by picosecond illumination.[26] (See also Leigh, J. S. and Dutton, P. L. (1974), *Biochim. Biophys. Acta*, **357**, 67–77.).

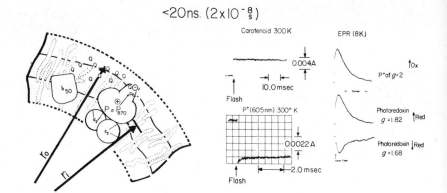

FIGURE 10. Q-switched ruby laser excitation of carotenoid band shift in R. spheroides chromatophores (Jackson, J. B. and Dutton, P. L. (1973), *Biochim. Biophys. Acta.*, **325**, 102–113; Dutton, P. L., Leigh, J. S. and Reed, D. W. (1973), *Biochim. Biophys. Acta.*, **292**, 654–664.

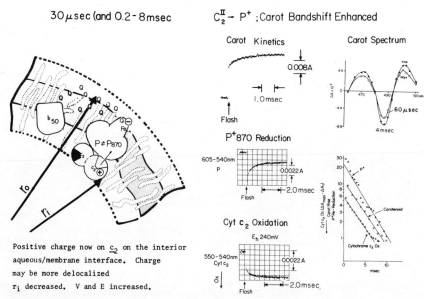

FIGURE 11. The enhancement of the carotenoid band shift by the oxidation of cytochrome c (Jackson, J. B. and Dutton, P. L. (1973), *Biochim. Biophys. Acta.*, **325**, 102–113).

a pH indicator outside the vesicle, we can detect H^+ uptake to the extent of one H^+ per electron transferred, measured optically.[34, 35, 36]

The complement of reduced ubiquinone (there are about twenty molecules of UQ per reaction centre) either diffuses or acts as a "bucket brigade" to transfer electrons and protons to cytochrome b 1.5 msec thereafter, as indicated optically by the reduction of cytochrome b. Cytochrome b can then further transfer an electron and release a hydrogen ion on the interior of the vesicle, completing the electron flow. We have now been all the way around the cycle from the illuminated reaction centre, the interchlorophyll charge separation, and then separation of charge towards the acceptor and thence towards cytochrome c, with subsequent movement of electrons through the pathway that generates ATP.

It is possible—and according to DeVault[37] and Chance,[38] it seems logical—that the electron transfer pathway can be summarized by a cycle (Fig. 13) in which cytochrome b operates as a redox pump between ubiquinone and cytochrome c. These two steps could occur at the same oxidation-reduction potential if cytochrome b underwent a conformation change causing a transition to a different potential so that it would be oxidized or reduced in an equipotential step of maximal efficiency from a thermodynamic standpoint.

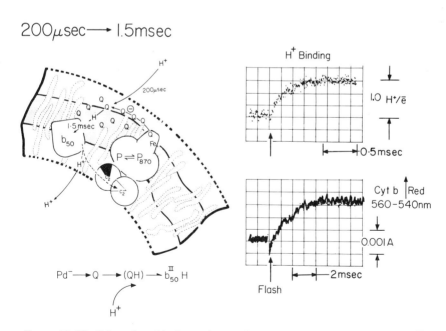

FIGURE 12. The light activated hydrogen ion uptake by the chromatophore membranes.[36]

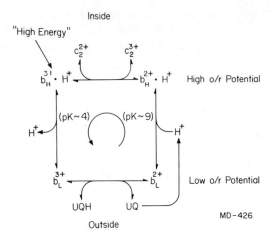

FIGURE 13. An energy conserving pathway for the interaction of ubiquinone in cytochrome c via a cytochrome b of alternating and mid-potential.[14, 37]

A hydrogen ion may also move across the membrane and may be released inside. If cytochrome b is oxidized before the H^+ is released, it has both a dissociable H^+ and whatever structural energy it has accumulated in this cycle, giving it the possibility to couple this energy *directly* to the ATP-synthesizing enzyme, or *indirectly* by injecting a proton on the inside of the vesicle. Thermodynamics may require that both these processes by involved.

The time at which this H^+ is released inside the membrane is not indicated by the carotenoids, as is appropriate for the transmembrane proton transfer being an overall "initial" act. For this reason, we have put into the membrane an extrinsic probe which may tell us more about the ion release reaction and events occurring between cytochromes b and c. Various merocyanine dyes have been used for studying electric potentials in squid axons by Dr. L. Cohen.[39] One that we have used is a dianilino-bis-isoxalone (Fig. 14) which is neutral and has symmetrical groups on each end. It therefore has, according to Platt's theory[20] and experimentally as well (B. Chance, unpublished observations)—less solvent response but shows a high degree of isomerization when an electric potential is put across it.

A related probe studied by Bücher et al.[21] shows a shift of its major absorption band towards the red when it is put into a fatty acid multilayer and a potential is applied across the multilayer. Figure 15 shows that the red shift occurs clearly in the quadratic or double frequency response (— · —) while the linear response (– – –) involves very little red shift.

We have also put these merocyanine probes into chromatophores [40, 41, 42] where they bind readily to the lipid phase, to see if there is a red

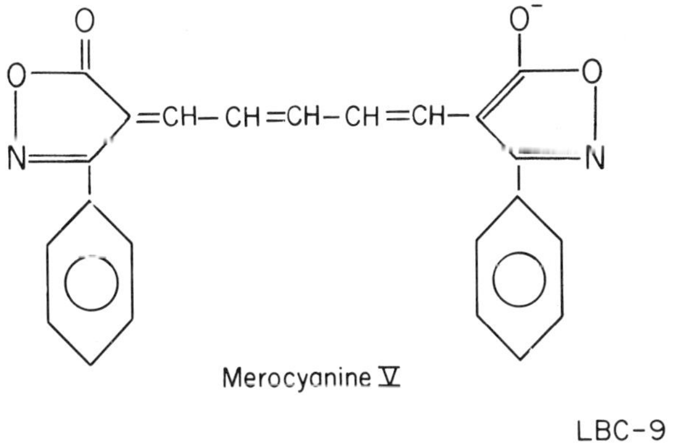

SPECTROSCOPY WITHIN THE CELL

FIGURE 14. Preliminary formula of MC-V (Russ, Stern, Denny, Cooperman, manuscript in preparation).

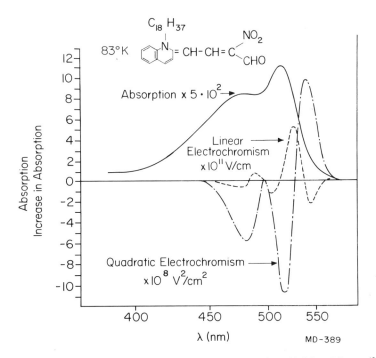

FIGURE 15. Linear and quadratic responses of a cyanine dye in artificial multilayers.[21]

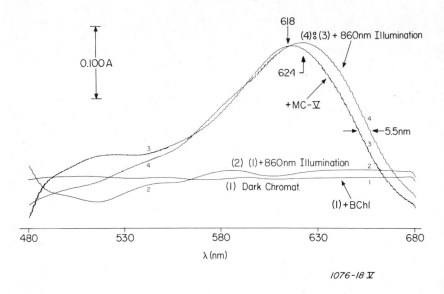

FIGURE 16. Absorption spectra of merocyanine probe in chromatophore membranes; response to illumination.

shift on illumination. Trace no. 3 of Fig. 16 shows the dark, merocyanine-saturated chromatophore. On illumination, the probe spectrum shifts 5.5 nm to a longer wavelength, suggesting that charge separation across the membrane has occurred.

This phenomenon is more readily identified with Bücher's result when presented as a difference spectrum as in Fig. 17, where the increased long-wavelength absorption at 643 nm and the decreased short-wavelength absorption at 603 nm are clearly demonstrated with respect to a dark or dimly illuminated specimen. This difference spectrum confirms the effectiveness of merocyanine-V as an electrochromic indicator.

The question remains as to when in the sequence of events that we have so far identified does the delocalized transmembrane potential appear. Figure 18 shows red shifts as a function of time after flash illumination. It can readily be seen that 50 msec is an appropriate half-time for generation of the electric potential. Thus, at the location where the voltage-sensitive merocyanine is located, we find none of the local charge separation events which are indicated by the carotenoids, but only the final consequences of membrane energization.

We have now covered the gamut of events of bacterial photosynthesis from the initial state of the membrane with the reaction centre reduced in the dark to, first, a 5 psec excitation interval and a charge separation in the

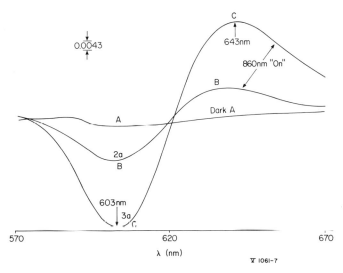

FIGURE 17. Absorption difference spectra of merocyanine probe in chromatophore membranes; light-dark difference spectra.

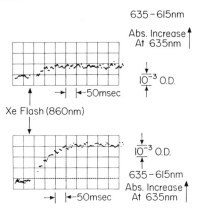

FIGURE 18. The speed of response of merocyanine probe absorbancy change to single flash in chromatophore membranes.

reaction centre. Figure 19 shows a more three-dimensional view of the membrane at this time, more or less as it would be with the lipids and carotenoids in the reaction centre.

The system is now prepared for the variety of electron transfer reactions that lead to the structure shown in Fig. 20, where the electron and the proton have both been transferred to cytochrome b via the ubiquinone system, forming protonated cytochrome b. Cytochrome c is oxidized and will not want the proton following its reduction. This we believe to be the energy-

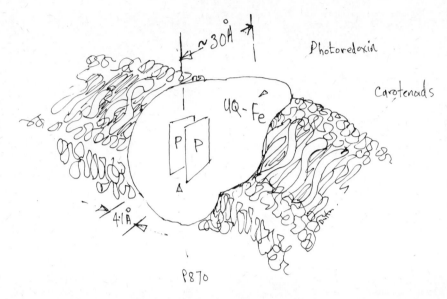

FIGURE 19. Drawing of possible membrane localization of components of the photosynthetic reaction centre protein in relation to the membrane structure (P. L. Dutton, unpublished data).

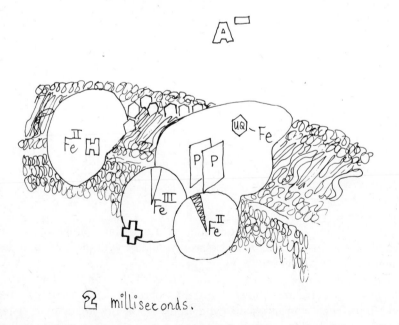

FIGURE 20. The electron transfer reactions which follow photoactivation of the membrane bound photosynthetic electron transfer systems (P. L. Dutton, unpublished data).

yielding step that gives the coupling reaction leading to ATP formation, and the step to which the merocyanine dye in the lipid phase of the membrane responds.

SUMMARY

This contribution has not cited support in depth for the multiple steps of electron transfer or for the details of energy coupling; reviews in this field are available. Nor has it dealt with opposing views which might be derived from the same data. Basically, however, the two theories for energy conservation—the "chemical" and the "chemiosmotic"—differ only in the relative significance of the proton ejected and the electron transferred in the energy coupling reaction. The chemical theory focuses on molecular interaction and uses as a model haemoglobin, where structural changes can result in large changes of dissociation constant, while the chemiosmotic theory puts all the onus for energy storage and conversion on the release of the hydrogen ion on the inside of the vesicle. The coupling between the transmembrane hydrogen ion gradient and potential and the ATP synthesizing system may be operating through both pathways.

Meanwhile, high resolution spectroscopic techniques ranging from X-rays for the location of the reaction centre and the cytochromes in the membrane, through NMR, EPR, etc., to picosecond laser optical methods, are being intensively and usefully applied to the problems of energy conversion in respiration and photosynthesis. Great progress has been made, but we still lack adequate tools for exploring the early steps of the excitation of the photosynthetic reaction centre for times shorter than 5 psec. We lack adequate sensitivity to apply more than optical spectroscopy to the observation of the kinetics of the light-induced electron transfer reactions. Obviously, diagrams and explanations such as these presented here must remain diffuse and controversial until the time course of structural changes of the electron carriers and of the membrane itself, atom by atom, are known as well as are the events of electron transfer.

REFERENCES

1. Warburg, O. (1949), "Heavy Metal Prosthetic Groups and Enzyme Action", The Clarendon Press, Oxford.
2. Meyerhof, O. and Kaplan, A. (1953), *Biochim. Biophys. Acta,* **12,** 121–127.
3. Krebs, H. A. (1948), *Harvey Lect.* **44,** 165–199.
4. MacMunn, C. A. (1886), *Phil. Trans. R. Soc.* **177,** 267–298.
5. Keilin, D. (1966), "A History of Cell Respiration and Cytochrome", Cambridge University Press, Cambridge.
6a. Hill, R. (1936), *Proc. R. Soc.,* **120B,** 472–483.
6b. Hill, R. and Hartree, E. F. (1953), *A. Rev. Pl. Physiol.,* **4,** 115.

7. Duysens, L. N. M. (1952), Ph.D. Thesis, Utrecht.
8. Lundegårdh, H. (1951), *Arch. Chem.*, **3**, 69.
9. Chance, B. (1951), *Rev. Scient. Instrum.*, **22**, 619.
10. Millikan, G. A. (1936), *Proc. R. Soc.*, **818B**, 366–388.
11. Fawcett, D. W. (1966), *in* "The Cell An Atlas of Fine Structure", (W. B. Saunders, ed.), Philadelphia.
12. Clark, M. (1928), *Hyg. Lab. Bull.*, **151**, 11.
13. Hill, R. (1956), *in* "Modern Methods of Plant Analysis, Vol. I" (K. Paede and M. V. Tracey, eds), pp 393–414, Springer-Verlag, Berlin.
14. Dutton, P. L. and Wilson, D. F. (1974), *Biochim. Biophys. Acta*, **346**, 165–212.
15. Witt, H. T. (1971), *Quart. Rev. Biophys.*, **4**, 365–479.
16. Jackson, J. B. and Crofts, A. R. (1969), *FEBS Lett.*, **4**, 185–189.
17. Kornberg, H. L. (1966), *Biochem. J.*, **99**, 1.
18. Smith, L. and Ramírez, J. (1959), *Arch. Biochem. Biophys.*, **79**, 233.
19. Smith, L. and Baltscheffsky, M. (1959), *J. Biol. Chem.*, **234**, 1575.
20. Platt, J. R. (1961), *J. Chem. Phys.*, **34**, 862–863.
21. Bücher, H. (1969), *Chem. Phys. Lett.*, **3**, 508–511.
22. Schmidt, S. and Reich, R. (1972), *Ber. Bunsenges. Phys. Chem.*, **76**, 589, 599, 1202.
23. Kamen, M. D. (1963), "Primary Processes in Photosynthesis", Academic Press, New York.
24. Netzel, T. L., Rentzepis, P. M. and Leigh, J. S. (1973), *Science*, **182**, 238–242.
25. Kaufmann, K., Dutton, P. L., Netzel, T. L., Leigh, J. S. and Rentzepis, P. M. (1975), *Science*, **188**, 1301–1304.
26. Leigh, J. S., Netzel, T. L., Dutton, P. L. and Rentzepis, P. M. (1974), *FEBS Lett.*, **48**, 136–140.
27. Blasie, J. K., Torriani, I. and Dutton, P. L. (1974), *Biochemistry/Biophysics 1974 Meeting Abstr.*, Minneapolis.
28. Dutton, P. L. and Leigh, J. S. (1973), *Biochim. Biophys. Acta.*, **314**, 178–190.
29. Okamura, M. Y., Moskwitz, E., McElroy, J. D. and Feher, G. (1973), *17th A. Mtg. Biophys. Soc.*, Columbus, Abstr. FPM-B2.
30. Parson, W. W. and Cogdell, R. J. (1975), *Biochim. Biophys. Acta.*, **416**, 105–149.
31. Norris, J. J., Uphaus, R. A., Crespi, H. L. and Katx, J. J. (1971), *Proc. Natn. Acad. Sci., U.S.A.*, **68**, 625–628.
32. Kaiser, W. (1974), *Appl. Phys. Lett.*, **25**, 87–89.
33. Sistrom, W. R. and Clayton, R. K. (1964), *Biochim. Biophys. Acta.*, **88**, 61–73.
34. Chance, B., Crofts, A. R., Nishimura, M. and Price, B. (1970), *Europ. J. Biochem.*, **13**, 364.
35. Cogdell, R. J., Jackson, J. B. and Crofts, A. R. (1973), *Bioenergetics*, **4**, 211–227.
36. Petty, K. M. and Dutton, P. L., *Arch. Biochem. Biophys.*, in press.
37. DeVault, D. (1971), *Biochim. Biophys. Acta.*, **226**, 193–199.
38. Chance, B. (1974), *in* "Dynamics of Energy Transducing Membranes" (L. Ernster, R. W. Estabrook and E. C. Slater, eds), pp. 553–578, Elsevier Scientific Publishing Co., Amsterdam.
39. Davila, H. V., Cohen, L. B., Salzberg, B. M. and Shrivastav, B. B. (1974), *J. Memb. Biol.*, **15**, 29.
40. Chance, B. and Baltscheffsky, M., with Appendix by W. W. Cheng. (1975), *in* "Biomembranes", (L. Manson, ed.), pp. 33–59, Plenum Publishing Corp., New York.
41. Chance, B. (1975), *in* "Biochemistry Series: Energy Transducing Mechanisms (E. Racker, ed.), pp. 1–29, Butterworth & Co., Ltd., London.
42. Chance, B., Baltscheffsky, M., Vanderkooi, J. M. and Cheng, W. W. (1974), *in* "Perspectives in Membrane Biology" (S. Estrada-O and C. Gitler, eds), pp. 329–369, Academic Press, New York.

Chapter 12

Computer Techniques for Spectroscopy

by E. J. Millett, Mullard Research Laboratories, Redhill, Surrey, England.

INTRODUCTION

Experience with the on-line reduction of data for the spark-source mass-spectrographic analysis of solids[1] led us to explore the feasibility of extending these techniques to other spectroscopic and related measurements. We designed a multipurpose multiple input system to share computing power among several experimental systems ranging from simple curve fitting for electronic lifetime measurements, through nondispersive X-ray spectrometry to fast signal averaging. This paper will explore the factors affecting the performance of computer systems for laboratory instrumentation in general, and discuss the scope and limitations of our own and commercially available systems.

Discussion of computer systems is complicated by the very wide range of system costs and computing power on the one hand and the differences in response times and computing loads met in practical applications on the other. The main factors determining the computing power of a system are the speed and efficiency of the computer itself and the speed and capacity of the peripheral and interface hardware. The computing load generated will depend on the inherent complexity of the application; the operating demands such as repetition rates, interaction with other systems, and finally the software strategy adopted to carry out the computation.

Taking all these factors together the Computer, Hardware, Application Operation and Software, gives an apt acronym for the subject (CHAOS). To introduce some degree of order, I propose first to discuss the choice of computer, then the information content and computing loads generated by some typical applications and finally computer systems and interface structures and their ability to deal with the data flow and computation required.

The demand for computer data sampling in spectroscopy is dominated by applications in analytical chemistry where the routine examination of large numbers of samples or the manipulation of complex spectra from relatively expensive capital equipment produces a large work load. While the examples used are mainly from this area they should enable us to deduce the characteristics of systems required to serve a wider range of applications.

COMPUTING LOADS IN PRACTICE

Computer calculation speeds

The immediate response required in instrumentation applications is virtually incompatible with conventional large shared multiple-access machines so that we are usually dealing with special purpose systems devoted to driving one or a few instruments. Practical economics dictates that the computer investment has to be comparable with the price of the instruments involved so that apart from some major nucleonic and astronomical installations most laboratory systems will be based on a "minicomputer". The definition of a minicomputer is elastic but as we shall need to discuss systems including major peripherals such as magnetic tape or disk drives we are talking about systems currently costing up to £100,000, but with central processors commonly using a 16 bit word and with 32 K word core memory or less.

The calculation speed of any computer is ultimately limited by the speed of the devices used in the logic. TTL integrated circuit logic with response times in the 10 μsec range represents the norm that is likely to be with us for some time to come. On this basis the computer cycle time which involves several logical steps is generally going to fall in the 0.2 to 1 μsec range. Slower "economy model" computers may remain, but the basic logic speeds are unlikely to increase significantly for the next decade.

The corresponding calculation times (assuming that floating point numbers are manipulated by software), broadly representative of the performance of current computers, are collected in Fig. 1. Floating point hardware is commonly available with the latest minicomputers and will increase the speed of most calculations by about a factor of ten. Nevertheless, it is clear that the evaluation of simple expressions will normally take 10 msec or so and operations such as curve fitting may occupy anything from a second to several hours.

We can assume therefore that the choice of central processor has little effect on the basic computation time, so that we have only to look at the

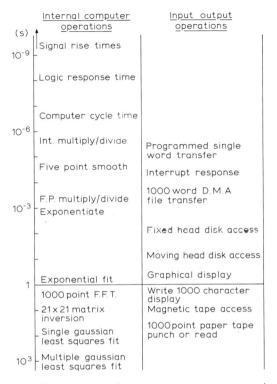

FIGURE 1. Time constants of some common computer operations.

amount of calculation and the frequency at which it is required to estimate the computing demand generated in each application.

Scanned line spectra

Simple line spectra and sequential control

The X-ray fluorescence spectrograph has the simplest possible physical characteristics. Each line corresponds uniquely with an element (Fig. 2). Digital measurement is inherent in the technique and computer data handling is only added to facilitate the sequence of successive measurements needed for the analysis of complex materials. In commercial systems of this kind the output may be on a simple printer and the computer is not only dedicated to the instrument but has a fixed program and is virtually inaccessible to the user. While the control computer may be severely under-used, perhaps operating usefully for only about 0.1% of the time, the system is not capable of expansion to deal with more complex data manipulation (for example, calculated fluorescence and absorption corrections) and probably never should be. Built-in dedicated control of this kind is more likely to be carried

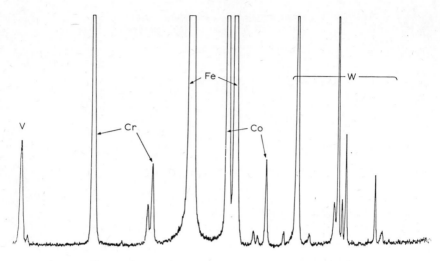

FIGURE 2. X-ray fluorescence spectrum of a stainless steel.

out in the future using systems based on microprocessors, and will become a standard feature of most laboratory instruments for repetitive fixed tasks.

Complex line spectra

In spark-source mass spectrometry the isotopic pattern required to identify an element may be relatively complex and subject to interferences (Fig. 3). Each spectrum with multicharged and molecular ions may contain up to 500 lines. Photographic recording with multiple exposures (10 or more) has to be used to give the required dynamic range for a complete analysis. This is an obvious candidate for automation and the techniques we have used are similar to those adopted for high-resolution organic mass spectrography[2] and optical emission spectrography.[3]

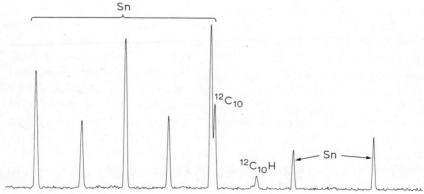

FIGURE 3. Section of a spark source mass spectrum showing hydrocarbon interference lines.

Conversion of the photographic record into digital form involves rescanning the spectrum with a digital microdensitometer. A 25 cm plate measured every 5 μm gives 50,000 data points for each spectrum. This is too large an array to hold and manipulate in core, so whether held in backing store or not the data must be scanned a section at a time by the software. It seems a natural consequence to avoid the use of backing storage by scanning the data within the computer while the spectrum is being physically scanned by the microdensitometer—real-time on-line operation.

Lines can be located by noting the points at which the first differential of the raw data passes through a preset critical value. A simple set of numerical coefficients will give the first differential of the least squares fit quadratic through five points[4] and still leave all values within the fast integer arithmetic range for a digital conversion range of 1000 to 1. This running five-point smoothed differential forms the first stage of a processing loop that can be fed raw data continuously to produce a list of line positions and intensi-

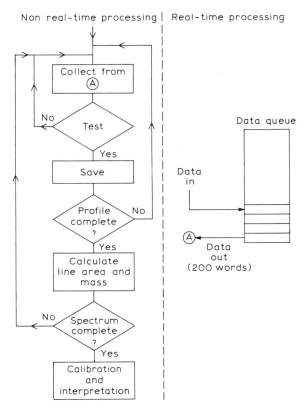

FIGURE 4. Buffered real-time data reduction program.

ties (Fig. 4). The problems of synchronizing data collection and processing are simply avoided by using the computer-interrupt procedure and some data buffers. Data transfer is autonomously initiated by the densitometer approximately every 5 μsec under interrupt and takes about 0.1 μsec or 2% of the processor time. A foreground buffer of 200 words allows up to 1 second processing time before overflow occurs, when the densitometer scan would be stopped.

The longest processing time in the background program is taken up by the intermittent calculation of the line intensity, integration and peak position interpolation.

Using the Hall function the expression

$$I = d \left| \sum_{i=1}^{n} \left(\frac{10^{(A_i - A_f)} - 1}{1 - 10^{(A_i - A_f)}/10^{A_s}} \right)^{1/G} - n \left(\frac{10^{(A_b - A_f)} - 1}{1 - 10^{(A_b - A_f)}/10^{A_s}} \right)^{1/G} \right|$$

has to be evaluated for each line. (A_i, A_b, A_f are measured line, local background and plate fog absorbences, n is the number of points in the profile and G, d and A_s are calibration coefficients.) Nevertheless the average time taken is no more than 50 msec per line or about 10% of the processor time during a 4 minute scan. Once the line list has been completed the task of quantifying the line intensities and identifying the elements present is carried out by a separate program.

This procedure has been described in some detail because it introduces some features of on-line data processing that are common in many applications. Digitization may first generate a very large data set of undefined length, but it is possible to reduce this to a simple list of position and intensity at high speed. The subsequent data manipulation required for qualitative and quantitative interpretation is then run intermittently between spectral scans. The processing is then clearly divisible into three phases, the fast (millisecond) response required for data acquisition, the data reduction carried out between spectral lines in a fraction of a second, and finally the data manipulation and display between spectral runs which can occupy any time acceptable to the operator, often several seconds or more.

Having produced a line list the computing load and timing is then entirely dependent upon the operating demands of the experiment. An illuminating example is the work of Klaus Biemann on the emergency diagnosis of drug overdose conditions in comatose patients.[5] A gas chromatograph separates the components of a blood extract into a series of peaks containing one or two compounds only. A mass spectrometer coupled to the gas chromatograph samples the gas stream every four seconds and generates spectra which are reduced to line lists in real time and compared with a limited library of probable compounds to provide positive identification of the constituents of

```
Palmitic acid
 Palmitic acid
  Palmitic acid
   Palmitic acid
       Palmitic acid
                  Oleic acid
                      Oleic acid
                          Oleic acid
                             Oleic acid
                    Oleic acid
                Oleic acid
              Oleic acid
              Oleic acid
                Oleic acid
                  Oleic acid
                     Oleic acid
                         Phenobarbital
                         Phenobarbital
              Phenobarbital
     Talbutal
       Oleic acid
        Oleic acid
      Oleic acid
        Oleic acid
         Oleic acid
         Oleic acid
```

FIGURE 5. On-line output of a coupled gas chromatographic mass spectrometer system. (After Biemann[5]).

the unresolved chromatographic peaks. The print out delineates the chromatogram and names the compounds found. (See Fig. 5).

The spectral identification program is run intermittently in the intervals between spectral scans and the print-out of the compound names occurs while the chromatogram is being run, which lasts several minutes. To the operator the information appears to be almost synchronized with the relatively slow chromatogram with only a few seconds delay.

Multichannel and transform spectroscopy

In some forms of spectroscopy the signal is not simply scanned sequentially but is recorded in parallel (as in multichannel γ-ray and energy dispersive X-ray spectrometry) or successive scans are repetitively added to increase the signal-to-noise ratio. The raw data input rate may be very high, perhaps exceeding that of a conventional computer and therefore demanding fast preprocessing, and a section of memory has to be dedicated to storing the spectrum. In the various types of transform spectrometry such as Hadamard[6] and Fourier transform[7] spectrometry the additional computation required between spectral runs to convert the stored information into the usable format of a scanned spectrum not only needs a large amount of core store but imposes fairly lengthy computation times.

Subsequent processing in both cases can follow the same general pattern as for line spectra but the computer system used may already be fairly heavily loaded at both the data reduction and data manipulation levels, so that in general these applications may merit a separate dedicated processor.

Band spectra

Complex spectra containing numerous lines and bands such as molecular infrared absorption spectra cannot readily be reduced to simple line lists like the inorganic line spectra. While various attempts have been made to reduce the information content of such spectra[8] and increase the efficiency of file searching by encoding schemes[9] these have not yet made sufficient impact on the problem to be generally accepted. The newer techniques of pattern recognition[10] may contribute to both the allocation of specific spectral features to related molecular structures and to spectral identification, but are still in the proving stage. The task of comparing a single complex fingerprint spectrum with any significantly extended file or library of recorded spectra remains a time consuming task even with computer help.

Superficially simple tasks such as the resolution of multiple overlapping peaks using least squares fit procedures will not necessarily converge to a satisfactory answer. Again such tasks require extended computing time and operator interaction.

Spectral interpretation and manipulation therefore will sometimes provide computing loads in which the computing time is long compared with the human operators response time or ability to absorb and act on the results, so that processing of this kind takes us outside the time constants of real-time response.

To sum up we can generally recognize four main phases which can be dealt with in the computing time available between operations in a real-time instrumentation system. They are:
(1) initial signal sampling and data transfer;
(2) data reduction instrument control and display, carried out "simultaneously" with data collection (Fig. 4);
(3) intermittent data manipulation with display, print-out, or plotting of results between spectral runs (Fig. 5);
(4) interactive sequence control using programmed or manual adjustment of instrumental parameters from run to run on the basis of the experimental results obtained in each run.

Control usually involves relatively few input or output operations at infrequent intervals, so it commonly represents a trivial part of the computer load compared with the data handling. The timing problems are otherwise analogous to those of data manipulation so control is not considered as a separate topic.

COMPUTER SYSTEMS

In an on-line system it is not only the computation times that matter but the time taken to gain access to the appropriate program in the system. This is more critical for the first two phases, data acquisition and data reduction, where the timing has to match the data input rates, than for data manipulation and display which only have to match the relatively slow response times of the human operator.

Where the computing load is large enough to require the use of additional memory such as magnetic disk or tape the access times to these peripherals, which lie in the range tens of milliseconds to tens of seconds (Fig. 1) have to be taken into account in the system strategy.

A minicomputer supplemented by such additional back-up store becomes quite a powerful computer, and it is possible to share its resources among several instruments, but in shared instrumentation systems the "interface" between computer and instrument may be realized in many different ways which interact with the computer system strategy.

To begin with, the term "interface" is somewhat misleading because it implies the interaction between only two systems. In practice there are several levels of interfacing. Analogue signals may have to be amplified and filtered, then digitized and sampled and finally entered into the computer input/output system in synchronism with the computer word cycle. These logical operations may themselves be distributed geographically and interconnected in different ways, and diverse signal transmission techniques may be used. These will all affect the timing and organization of data transfer in the system. The relative merits of the different combinations of computer system and interface structure are best discussed by examining the performance and limitations of some typical examples.

The Dedicated Minicomputer

In a simple dedicated minicomputer system the individual analogue and digital signal lines can be brought in directly to standard interface units plugged into the computer input/output bus (See Fig. 6). Such a system is usually tailormade for the application and the program may be written entirely in assembly code to pack it into the smallest possible core space. This is typical of the approach used by equipment manufacturers supplying N.M.R., Mass Spec., I.R., X-ray diffraction and similar instruments where the effort involved in designing the hardware and software can be recouped by sales to many customers with a common basic requirement. With the reducing cost of minicomputers the need for close packed programs is declining and most research laboratories will want some extension or modification of the standard program. However, without additional peripherals program editing is a very slow and cumbersome task so that the simple dedicated system offers the least program flexibility. It has the advantage that data can be entered under program control or using direct memory access at peak rates approaching the cycle time of the computer (i.e. up to

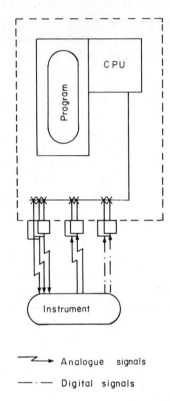

Figure 6. Simple dedicated computer system.

10^6 words per second). This is not readily compatible with simultaneous data entry from other systems and is the preferred approach for parallel recorded spectra. With additional front end preprocessing hardware, dedicated systems can deal with the fastest multichannel analysis and signal averaging problems.

A minicomputer complete with magnetic tape and disk store displays and printers, may still be used in a dedicated system, but systems of this size usually require a team of full-time electronic engineers and programmers and are only found in major installations such as nuclear accelerators, radiotelescopes, etc. Flexibility is only achieved by the use of skilled personnel who may even write their own high level programming language for the particular scientific community.[11]

Multiple Access Computer Systems

A number of authors have described the operation of several instruments simultaneously using a dedicated minicomputer and direct interfacing of

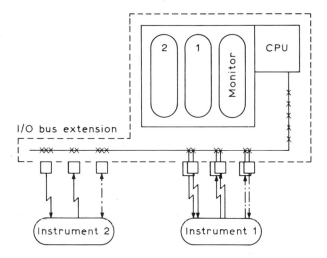

FIGURE 7. Multiple input dedicated system.

each instrument output to the computer (Fig. 7). Program access is kept fast and core space minimized by using a special purpose "monitor" to control the system. This has the same general advantages and disadvantages as the single instrument dedicated computer, but may become unduly complicated with many instruments, particularly if they are separated by any great distance, when the cost of cables and the complications of noise pick-up on analogue signals become significant.

Another approach uses the existing systems for process control which can be adapted for the purpose. For example, the conventional time shared multiplexing units which take data at fixed intervals of say 100 milliseconds are used for data capture together with one or more fast channels for instruments outside this range (Fig. 8). Such systems are commonly run using the standard time sharing executive with data handling programs held in reserve on a fast disk store, and real-time programs limited to data acquisition. They still have long multiple signal lines to distant instruments and offer little scope for real-time data reduction and control because of the timing limitations in gaining access to the stored programs through the executive.

The range of data rates that can be handled can be increased by using one of the commercial process control interfaces which can be purchased with almost any defined array of digital and analogue inputs and outputs, but these do not overcome the problem of long multiple connections to distant instruments. This can be reduced by using a local instrument interface to handle the collection of signals required at the instrument and twisted-pair digital cables for connection to the computer I/O interfaces as in the Varian

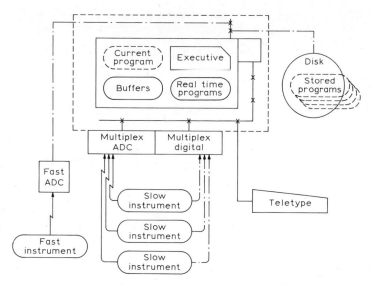

FIGURE 8. Time shared multiplex system.

I.B.M. "Spectroshare" systems (Fig. 9). This procedure separates the instrument signal interface from the computer interface and has the added logical advantage that each instrument is connected to the computer at a single point.

In a research environment the array of signals needed to run any one instrument may not conform to the standard pattern, and may be subject to change as experimental conditions require. To meet this need flexible and adaptable modular interface systems are the only satisfactory answer. The "Camac" system designed originally to meet the demands of the atomic energy industries is the best known of these systems and naturally offers a wide range of fast input systems with local logic and storage to supplement the computer, complete with signal distribution systems for both local and distant apparatus. However, the use of a 24 bit word and the degree of complexity and built-in autonomy make this system relatively expensive and cumbersome.

The MRL Hierarchical System

In our own design we set out to provide a structure that would combine the best features of dedicated and shared computer systems, with a flexible and adaptable instrument interface. At the centre of the diagram (Fig. 10) we have a single real-time 16 bit 16 K word minicomputer with a local digital multiplexing interface. This is connected to the computer I/O system and

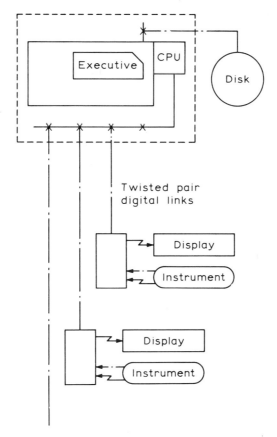

FIGURE 9. Remote digital instrument interfaces on a time shared system.

incorporates a very fast digital multiplexer scanning through 16 inputs in 4 microseconds. One of these is connected to the batch computer, to which we shall return, and the remaining fifteen are possible connections to remote instruments.

All the interconnections carry digital data only and are made with coaxial cables using a parallel to serial conversion and back. These links are capable of carrying data in both directions simultaneously so that a single coaxial cable carries all input and output to and from each remote instrument. Such a cable can be driven at high frequencies (5 MHz or 160 000 16 bit words per second) independent of distance up to 500 metres. At each instrument a similar high-speed digital multiplexer scans up to fifteen instrument signal channels, and an array of analogue to digital conversion and digital input and output modules can be plugged in to match the particular instrument.

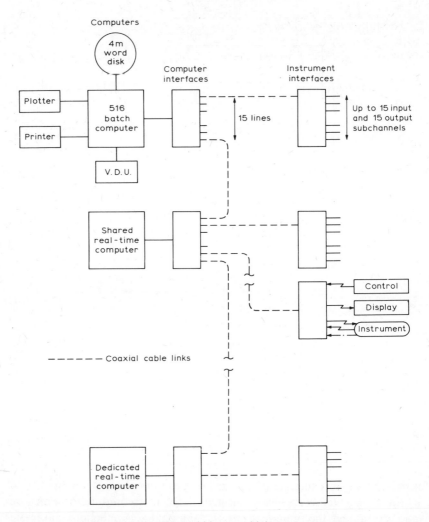

FIGURE 10. Adaptable multi-input multiple computer system.

The digital multiplexers impose a simple queueing system, each instrument is served in turn one signal channel at a time. In this way data from a slow source is automatically interleaved between inputs from a fast source without any special synchronization. However the scan times of the multiplexers are so short compared with the time taken by the computer to handle an input (a typical interrupt will take 60 microseconds) that this remains the time-limiting step. For peak data input rates below 5000 words per second the system will continue to serve each user as if he had a dedicated computer with only a slightly increased cycle time.

The programs required for real-time data reduction are all held in core with a simple executive designed to run programs in sequence while data transfers are carried out simultaneously under interrupt. Most programs run for only a fraction of a second.

In practice it is surprising how little program space is required for many applications and our current load runs from 7 K words for the energy dispersive X-ray analyser, complete with stored spectra and element identification, to 200 words for an integration. This is primarily because the programs can all share the basic maths library, input/output and display routines which occupy about 4 K of core.

Using the shared system and its flexible interface, small experiments that would not merit a dedicated processor (even at today's prices) can be set up economically and quickly for virtually the cost of the interface alone, so opening up computer data handling for a whole new range of applications.

Program generation can only be done rapidly using the power of a computer with back-up store for compilers, libraries, etc. With a dedicated or a single shared computer the real-time work must be interrupted for program generation. In our system this work and the temporary connection of real-time experiments for design and debugging is catered for by the batch computer fitted with the same fast 16 channel digital multiplexer. The link between the batch computer and the real-time computer allows the program configuration in the real-time computer to be "piped-in" from the batch computer to add another experiment or change a program virtually instantaneously. This computer-to-computer connection also allows real-time reduced data to be transferred to the batch machine for interactive operations or file comparison, so catering for non-real-time processing. Finally, the high-speed data handling tasks of the signal averaging and transform spectroscopy type can be tied in to the system using a remote dedicated computer. In our own system this included a fast multichannel analyser front-end for data rates in excess of computer speeds.

To sum up we have a system which has flexible and adaptable interfacing, with the system organized to separate the tasks of foreground signal processing, shared real-time data reduction, slow program development and data manipulation matching the tasks met in spectroscopy. While the costs of minicomputers will continue to fall the costs of software generation and peripherals are more likely to rise, so this hierarchical connection of computers will continue to offer economic and performance advantages over dedicated unsupported processors. The reliability of the minicomputer and its multiplexing interface is so high that it makes little impact on the user. Instruments may be connected or disconnected with impunity while the system is running so that so far as the user is concerned computing power comes on a coaxial socket.

In conclusion, it should be emphasized that this system is experimental and is not available commercially.

REFERENCES

1. Millett, E. J., Morice, J. A., and Clegg, J. B. (1974), *Int. J. Mass Spec. and Ion Phys.*, **13**, 1-24.
2. Venkataraghavan, R., McLafferty, F. W., and Amy, J. W., *Analyt. Chem.*, **39** (2), 178.
3. Kylstra, C. D., and Schneider, R. T. (1970), *Appl. Spectrosc.* **24** (1), 45.
4. Savitzky, A., and Golay, M. S. E. (1964), *Analyt. Chem.*, **36** (8), 1627.
5. Biemann, K. (1972), *in* "The Applications of Computer Techniques in Chemical Research", Hepple, Peter ed., p. 4, Institute of Petroleum.
6. Plankey, F. W., Glenn, T. H., Hart, L. P., and Winefordner, S. D. (1974), *Analyt. Chem.*, **46** (8), 1000.
7. Cuthbert, S. (1974), *J. Phys. E., (Scientific Instruments)*, **7**, 328.
8. Rann, C. S. (1972), *Analyt. Chem.*, **44**, 1669.
9. Jurs, P. C. (1971), *Analyt. Chem.*, **43** (3), 364.
10. Kowalski, B. R., and Bender, C. F. (1973), *Analyt. Chem.*, **45**, (13), 2234.
11. Moore, C. H., and Rather, E. D. (1973), *Proc. I.E.E.E.*, **61**, 9, 1346.

Discussion

A discussion period was available after most of the papers and resulted in a lively exchange of questions and answers and some argument. These discussions were most valuable, but have not been reported in detail except where some new idea or concept emerged or where fuller explanation seemed necessary. Instead, the principal speakers were asked to include clarification and answers wherever possible in revised versions of their talks prepared for publication. Some of the points requiring further elucidation have been combined below in the account of the general discussion held on the second afternoon of the conference. In this way it is thought that most of the essential points have been covered.

Professor G. W. Series described a technique for artificially narrowing a linewidth using a modified Fourier transform of the time decay following pulse excitation. This involves throwing away a certain amount of unwanted information and caused a good deal of heated argument, a number of speakers maintaining that no real advantage is gained. Professor Series has contributed a written description of the method and has answered some of the criticisms raised against it. This is included as a short paper at the end of this discussion.

A good deal of discussion initiated by Professor R. Hill took place on the relationship between coherence in laser light and entropy. We include two of the contributions which help to clarify the situation.

The first is from Professor A. Kastler, as presented during the discussion, and the second subsequently supplied as a written contribution from Professor R. K. Bullough.

DISCUSSION

Professor A. Kastler: I may perhaps direct your attention to some effects of interaction between light and matter which can now be studied by using laser light. They are not nonlinear effects but just linear effects. I shall speak of what we may call non-resonant interactions between atoms and light. We know that atoms interact strongly with photons when the frequency corresponds to the energy difference of two atomic levels. We know also that an atom can interact with light of any frequency. We know this from the phenomena of refraction or dispersion which is related to refraction and scattering of light. All these effects are actions of the atoms on the light. The presence of the atoms changes the properties of the light. At the same time there is a reciprocal effect: the light changes the properties of the atoms. This has been demonstrated first by Cohen-Tannoudji using ordinary light sources. He has shown that by non-resonant light one can change or displace the energy levels of the atoms—an effect related to the Lamb shift, and that we can also change magnetic moments of atoms. One can even reverse the sign of magnetic moments. Of course all these effects can be highly enhanced by using laser light. It may now be possible to displace energy levels of atoms by shifts of several thousand MHz or still more.

Another aspect of the interaction of light and matter is in the phenomena of coherence. We have seen that laser light is very coherent and can be used especially to study coherence effects in atoms. May I just draw your attention to an old work which has been practically forgotten? It is the thermodynamic aspect of light coherence. When we have two incoherent light beams, then we know that the entropy of those beams together is equal to the sum of the entropies of the two light beams. The entropy is an extensive variable of the system and the theorem of additivity of entropy applies. But when one has two light beams which are mutually coherent, then this theorem of additivity does not apply. This was shown by Max van Lauer in two remarkable papers which were published in Annalen der Physik in 1906 and 1907. The coherence of two or more light beams is related to a certain neg-entropy. This is not surprising as coherent light is an ordered effect and incoherent light is disordered. To this disorder corresponds a definite neg-entropy, and I think this is very important. If you look on the laser you may see it is a system which generates order, neg-entropy, continuously, in generating coherent light. But as a laser beam expands it must at the same time generate entropy. Of course every such laser system is a dissipative system which generates much heat. We must give some outside energy to the system. Then the system also generates neg-entropy and, at the same time, entropy. I think it would be very interesting if this thermodynamic aspect of coherent light could be studied more. Life is also a generation of order—a multiplication of order. In a laser we have such a system. It multiplies physical order continuously, but it must be nourished at the expense of energy and neg-entropy, and

I think it would be very interesting to study further these relations.

Professor R. K. Bullough: I would like to comment on two suggestions made by Professor Kastler, and on some further questions concerning entropy.

First, both thermodynamical entropy and optical coherence can be viewed as measures of order and it is sensible to suggest that study of this connection will prove a fruitful one. However, such a study has been under way in the years since about 1968 most especially in Haken's group at Stuttgart and Scully's in Arizona. One of the more striking connections between coherent optics and thermodynamics is the rather complete analogy between the onset of laser action with increasing pumping and a one-dimensional phase transition. Providing the optical system satisfies detailed balance in the steady state one can define a generalized entropy $S = \ln P$ where P is a probability function usually satisfying a time independent Fokker-Planck equation. In this way it is possible to develop a theory of fluctuations close to the optical "phase transition" indistinguishable from Landau's phenomenological theory of phase transitions in thermodynamical systems.

Just as order appears in the superconducting or ferromagnetic phases by changes in the symmetry of the states, so order appears in the optical system by a change in the symmetry of the light field amplitudes with respect to their phase. Comparable far-from-equilibrium steady states occur in chemistry and Prigogine has called the order introduced in these "dissipative structures". There is a fairly complete analogy between Prigogine's theory of dissipative structures and the order theory of optical coherence, although it will surely give Dr. R. Hill some pleasure to learn that indeed the precise definition of entropy in this context is still a "mind scratching exercise" demanding further study. Graham (Stuttgart) treated this problem at the last Rochester Conference (on Coherence and Quantum Optics) and was able to treat, for example, optical subharmonic generation and third-order subharmonic generation from this point of view.

In my opinion, this line of research is important because it must be the way to treat (theoretically!) biological systems *in vivo*. These are certainly far from equilibrium steady states; and when they are interacting with light it may become a natural extension of the theory to treat the light this way as well.

The second suggestion by Professor Kastler allows me to comment on one of the methods of high resolution spectroscopy; saturated absorption spectroscopy, which appeared briefly in Professor Jacquinot's review. Of course, intense light shifts the energy levels of atoms and if the saturating intensity in a saturated absorption experiment were perhaps one order of magnitude higher one could worry about such shifts. A rough calculation shows that fields of about 1 Wcm^{-2} per MHz can shift atomic levels by amounts comparable with the hyperfine splittings. But the further analysis we presented at San Francisco in June shows two things. First, if the saturating field is a single-mode field, coherent or not, there is no light shift. Second, if the saturating field has a larger bandwidth but is coherent there will be no simple light shift (other things may happen such as the optical nutation described by Dr. Brewer).

In order to see such a light shift it is necessary that the fields are both incoherent and of fairly broad band. In a saturation experiment one could, in principle, impose an incoherent field in modes lying at an angle to the colinear probing and saturating fields. But these incoherent modes would have to have a bandwidth of the order of the natural width—meaning that the imposed incoherent field would have to have an intensity of about 1×10^3 Wcm^{-2} for shifts of 10^9 Hz. It would certainly be interesting to devise experiments to see such light shifts: for quantitative confirmation of theory at optical frequencies is still desirable because, for example, the part of the light shift due to "the square of the vector potential" should not function in the observed shift and the dependence of the shift on the distribution of mode occupations has also not been verified. Nevertheless, the conclusions of our analysis are: first, that such light shifts are reasonably well understood theoretically; second, that the intensities and coherence of saturating fields used in optical spectroscopy are such that we need not worry about the effect of these fields on the observed spectrum. It is interesting to speculate whether such light shifts have any significance to the investigation of biological or chemical systems with the more intense fields contained in ultrashort optical pulses, or indeed whether the state of coherence of these fields has any significance here.

The question was raised whether two-photon excitation would be applicable to molecules. Professors Jacquinot and Townes thought so, though there might be some difficulties with the multiplicity of the levels. Indeed, as Professor Townes pointed out, the vibrational states of molecules would be very favourable for this technique with suitable tunable infrared lasers, since the levels, being just slightly anharmonic, almost resonate for each photon. There should therefore be a favourable case for a large number of resonances.

Considerable discussion took place on whether any real advantage is gained by using a heterodyning technique. Dr. E. R. Pike questioned, for example, the advantage claimed by Dr. Brewer in experiments (three orders of magnitude). The result of the discussion seems to be that in the visible region of the spectrum, while one can obtain a large increase in signal there is not the same gain in signal-to-noise ratio, since most of the noise comes in with the signal. In the infrared, however, detector noise generally limits performance and here increase in signal also means increase in signal-to-noise ratio.

It was also pointed out that the heterodyning technique was an alternative to the photon counting techniques described by Dr. E. R. Pike in his paper and had been used very successfully by Professor J. B. Cumming and his colleagues at Columbia and also by Professor G. B. Benedek and his colleagues at MIT to study a number of the systems described by Dr. Pike, including blood flow and motion of large molecules. Dr. Pike claimed that when the fluctuations to be observed are slow, or velocity shifts are small, the photon-counting technique had decided advantages.

Dr. R. G. Brewer explained in discussion with Professor A. Kastler that the collisions he described, which do not disturb the phase, take place between molecules which have no angular or orbital momentum. These concern vibration–rotation transitions in the ground electronic state. Professor Kastler indicated that optical pumping experiments similarly show that S-states are unaffected by phase-interrupting collisions whereas P-states are.

Dr. Brewer also indicated that in his experiments on optical nutation a dynamic Stark effect had been observed. This appears as side-bands when the optical nutation signal is fed into a spectrum analyser.

Professor S. D. Smith raised the question of the usefulness of wavelengths around 5 µm as compared with the 10 µm region in infrared astronomy. Professor Townes was of the opinion that this is a very useful and potentially fruitful region to explore. The noise will be up by about a factor of two but there is some indication that the available signals are also up by about the same factor. This region is also rich in information due, amongst other things, to the molecule CO. He thought the development of tunable lasers in this region would be very valuable.

Considerable discussion took place on the limit set by photon noise in a detection system and there appeared to be some disagreement between Dr. E. R. Pike and Professor C. H. Townes on the advantages of a multiplex system. Finally this turned on whether one is observing one or more objects that can be resolved as point sources. Professor Townes was concerned with observing one point object (a star). He indicated that the multiplexing which is done through Fourier analysis puts all the signal on *one* detector. There are some systems that are being developed but not yet really profitably used, in which the spectrum is split up into many narrow bands and put separately into several detectors. This system, in principle, certainly will gain time on the telescope and in the long run could be quite important. It is similar to techniques which are used for military and industrial purposes but, when one is fighting the ultimate in sensitivity at the same time, the complication of many detectors is simply one which hasn't yet really been effectively put into practice.

Professor C. H. Townes: If one has separate points, then of course one can resolve the points. But my remarks were really directed at the single star—something that is not resolvable in the infrared. If one can resolve the separate points their statistical fluctuations can be separated. On the laser amplifier situation, the trouble is that it, too, in preserving phase, must have this same kind of fundamental noise and so amplifying may allow one to use a poorer detector but it doesn't in the long run change the absolute limit very much.

Some questions had been raised concerning the removal of doppler shift by using transitions involving two or more photons. Professor A. Kastler explained the basic physics of the process simply and this is now given.

Professor A. Kastler: I shall just make a short comment on one point of Professor Jacquinot's lecture—he had not time to explain it. It is the reason why the doppler broadening can be suppressed in two-photon transitions of the atom, or more generally in multiple-photon transitions. We have to go back to the origin of the doppler shift. We have an exchange of energy between the atom and the electromagnetic field. The atom with velocity will have kinetic energy $\frac{1}{2}Mv^2$ and also momentum $M\mathbf{v}$. A photon arrives with energy $h\nu$ and also a momentum $h\nu/c$. When the energy is transferred to the atom, there is always at the same time a transfer of momentum. The atom always has to take the momentum of the photon so that the momentum of the atom after the absorption of the photon will be $M\mathbf{v} + h\nu\mathbf{i}/c$ where $\mathbf{i}$ is a unit vector in the direction of the photon, and this changes the kinetic energy of the atom. This energy is taken from the photon energy. This is the origin of the doppler shift. If we have the atom interact not with one photon but with several photons, then the atom has to take over the resultant momentum. If this is zero there is no doppler shift. If we have a three-photon transition the track angles of the three photons can be such that the resultant momentum of the three photons is zero and the velocity of the atom will again not be changed. Of course to fulfil these conditions with two photons, they must come from opposite directions and have exactly the same momentum i.e. the same frequency.

The question was raised by Dr. Q. H. F. Vrehen as to whether there is any essential difference between a two-photon excitation and two sequential one-photon excitations, the energy sum of the photons corresponding to the same energies as the former two, provided no phase-interrupting collisions take place between the absorption of the first and second photon. Professor Jacquinot thought that there was then no essential difference. There certainly would be if such collisions took place between the absorption of the two photons. It may be remarked however, that in the second case we have *exact* resonance enhancement, while for two photons absorbed "simultaneously" this is not necessary, and indeed is usually undesirable because of the very strong absorption at exact resonance.

Some discussion took place on the use of up-conversion technique, e.g. raising an infrared frequency into the visible by mixing. Professor S. P. S. Porto described a new system being developed by a former student Martin Gunderson. A brief account is given below.

Professor S. P. S. Porto: A new system of up-conversion that seems to be extremely efficient, and looks quite interesting, has been developed. The idea is very simply the following: in a semiconductor one has a fundamental band gap and one may have in here other levels which are essentially excitonic levels. Some of these excitons are bound and quite easily ionized by a laser beam. Let us take for example gallium phosphide, with excitons related to impurities like bismuth which lie a few millielectron volts below the band gap. If one cools the crystal, one can reach this level easily with a laser, and then from here on upwards with an infrared beam. This up-conversion is broad-banded because any energy above the dissociation energy is absorbed. It is also extremely fast because of an interaction with the bottom of the band. So one shines something like, say, 647 Å light on the crystal and gets back something which is nearly in the yellow zone. This is a quantum counter, meaning that for each infrared photon absorbed one up-converted photon comes back. We had a water vapour laser working at 28 μm and 32 μm radiation, and one could see those things visibly. It is quite an interesting new up-conversion technique which I think has not been described. Besides being extremely fast it looks like being a very good detector.

If this technique could be extended to about 100 μm Professor Townes considered that it would provide a very valuable tool for astronomy.

The question as to what is known about the chemical composition of interstellar clouds was raised (by Professor Britton Chance). Professor Townes gave the following information.

Professor C. H. Townes: There are, by now, about 30 molecules that have been found in interstellar space—many of biological significance. In fact they are pretty much that collection of molecules which a biochemist interested in the origins of life would have said were the things he needed in order to imagine life getting started. They include a lot of formaldehyde, formic acid, recently dimethyl ether, methyl alcohol, various cyanide compounds and so on. Everybody in radioastronomy now believes that everything imaginable is there—it is just a matter of getting enough sensitivity and it will all be found.

Professor G. Stein made some comments on biological research as indicated below.

Professor G. Stein: Professor Chance's paper opened a field which some of us hope we shall hear more of under "Where do we go from here?". Some of these very highly refined methods will find a most interesting application in the biological field. If we are going to talk about that later, it seems to me that possibly other methods, such as infrared and Raman methods might be most valuable, and not only the methods that Professor Chance described as visible and ultraviolet. Professor Chance mentioned that one can—and one has always done so in biology—go to the full biological system, and even without understanding all of the component parts, try and apply new techniques and get most valuable information as he described. He mentioned the work of Wagborg and Kripps and so on, whereby taking the simpler parts of the biological system—chemical molecules which are still very complicated—say, enzymes or nucleic acids—and these separately, and by putting these together properly one can again gain from another point of view information valuable to the biological system.

We have done some work which I would like to mention which gives support to what Professor Chance has been doing, namely, take one component, cytochrome c which Professor Chance described in so much detail. In cytochrome c there is a very complex molecule yet which is much simpler than all photosynthetic apparatus. It has a metal part and has a protein part, and the method that one can apply here is the following. One uses a two-stage system, with a pulse which one can get either by pulse radiolysis or by laser catholosis tied down to the picosecond region. One then follows it up by using the kinetic spectroscopy which Professor Chance described. Here I would like to point out that fairly recently Professor Porto and Dr. West have published a detailed review article, most valuable, giving all the particulars of fast kinetic spectroscopy as applied in chemical and biological systems. We have, for example, used this two-stage method—fast pulse, followed by the kinetic spectroscopy as a standard technique, and indeed one can look inside the protein and follow the detailed steps of reactions. We have done this with cytochrome c and actually inside the protein one can follow the pathway and the time scale of the electromagnetic changes and also follow the spectroscopic changes. One or two of the points that I would like to mention are: there are some dangers in going down to the higher resolution for the following reason. One can of course readily get picosecond pulses but then one is faced with several technical difficulties arising from the fact that because of the short time the peak power is very high indeed. It was already indicated that it is absolutely mandatory to try and sort out one-photon processes from microphotonic processes. There is another thing. If the peak intensity is so high one can get inside new material nonlinear effects and one might get effects which are due to, say, the sum of two photons, so that one gets photochemistry which is not due to the line at 700 nm but twice the energy. So I think that down to the nanosecond region we are in pretty good shape technically in these biochemical and biological systems. I think the picosecond region offers very great challenges, but I think that quite extraordinary care should be applied in sorting out some of the phenomena which can arise because of the very high peak power.

Professor Chance indicated that while one had always to be careful with high power pulses since there are systems in which they can be damaging, nevertheless, this should not prevent their being used with care. In the systems he had described where one proton corresponds to one electron it isn't necessary to use large numbers of protons to define a suitable system.

Dr. Connerade also drew attention to the danger of damaging the sample with high peak powers. In X-ray experiments the damage depends on the time in which the energy is delivered to the sample. For light pulses the relationship between time and damage is not yet clear.

Professor G. T. Reynolds: In view of the above comments it might not be inappropriate to point out that we have applied a spectrometer in front of an image intensifier to study spectra *in vivo*. There are a few instances in which the in-vivo structures are somewhat different from the in-vitro structures. The purpose of the technique is to be able to record a complete spectrum from a single flash to a biluminescent organism. We propose to follow some of these suggestions that have been made here to put ahead of the present intensifier a sweep intensifier. Thus we can study the different wavelengths in these rather broader band emissions to see, particularly in those systems where we already have reason to think there are two protein systems, what the kinetic spectroscopy of the second protein would be.

Professor G. Weber: I would like to make a further comment on the applications of high resolution spectroscopy in biological systems. From the two kinds of high resolutions possible, spectral resolution and time resolution, spectral resolution is no use to us. We have no experiment in biology for which an available spectral resolution is inadequate. Time resolution we have found extraordinarily useful and what is the very high standard of resolution required by us? I think one can make a very good calculation of the order of magnitude. One can immediately see that we are interested in interruptions. To get an interruption between molecules one has to have a diffusion time and the diffusion time is, say, the shortest distance to require investigation in the order of 100 picoseconds. If one wants to be a little more optimistic, well, use 30 picoseconds.

Dr. Connerade: There is another example of high resolution spectroscopy applied to living tissues in recent experiments using synchroton radiation where very intense and collimated X-ray beams has enabled X-ray diffraction measurements on living tissues to be made in which actual changes in the tissues have been seen. For example in the retina, I think, of a frog when there is ambient light and when there isn't ambient light there is a change in the diffraction pattern which has been observed.

Sir Harold Thompson: Well of course this achieves high resolution in space. About 1949 I remember borrowing from Dr. Birch at Bristol that first reflecting microscope that he made, putting it in front of an infrared spectrometer and looking at a biological cell. The technique was such that we had a picture of the cell on a piece of white cardboard and we saw that some bits were black and some white and we moved the slit around and got different spectra for different parts. This might mean the different ratio of nucleic acids and sugars—or whatever it might be—but there were differences. This was simply dependent upon having an extremely microscopic beam of light focused on the different parts of the sample, so it was the same sort of thing one can do with an accelerator but here one uses soft X-rays.

A discussion led by Professor S. D. Smith followed concerning the possible use of infrared and Raman spectroscopy in the study of biological materials. The discussion concerned the type of information one might expect to get and, in particular, emphasized the possibility of studying vibrational structure of large molecules. Professor Stein summarized and commented as follows.

Professor G. Stein: I would not like to commit myself because I am not an expert on this but to me it seems that in biological systems you can look at a number of things. One of them is what Professor Chance described. You have the oxidation-reduction processes where gross change is visible in the ultraviolet spectrum. Another possible feature of biology which is very important is the configuration of changes. The energies involved in configuration changes are very much smaller than those we observe in visible and ultraviolet spectra and here it seems that when one gets a macromolecule and sufficiently change the situation so that certain very specific vibrations are affected because of the environment, I would say that Raman and infrared can be applied. Here we have situations where certain selected very well-defined vibrations—which are specifically selected—are affected in a dynamic form where something changes and one might observe small changes in infrared vibrational spectra, Raman frequencies and so on. I think there is perhaps quite a lot to be done.

DISCUSSION

On the more general question of the use of Raman spectroscopy for the study of solutions, Sir Harold Thompson commented as follows.

Sir Harold Thompson: I haven't entered into the Raman discussion although I have measured many hundreds of Raman spectra in my time. I was a little anxious earlier on when some were talking rather easily about the success of Raman work in biological systems. The big hope of Raman spectroscopy hasn't really worked out. It was that it would be able to be used widely in academia and in industry for the study of aqueous solutions. Now for reasons which have not yet been properly explained, it has not worked as hoped. We all know that Professor Lord at MIT is measuring aqueous solutions and a lot of other things—some of which were mentioned this morning—and he has measured some very fine spectra but he was the first to tell me that he couldn't understand a lot of the curious things that happened when these were measured and the differences which are still inexplicable!

Again, even though the laser has made things easier there are some problems which are recognized by people who have been measuring Raman spectra for the last 30 years which cause them some hesitation. I must express my own doubts about the use of Raman spectroscopy in these fields.

Professor Porto was more optimistic and particularly regarding Raman spectroscopy of solids in which great progress has been made in the past five years with the help of laser sources. He was of the opinion that some of the new results that are not understood arise not because of faults in the methods, but because of incomplete understanding of the physics.

Dr. A. Mooradian felt that some of the difficulties were semantic, physicists, chemists and biologists speaking different languages. Some were also caused by inadequate techniques, by poor signal-to-noise ratios.

Dr. A. Mooradian: When one studies the fluorescence from aqueous solutions, one finds that a salt, even in the most minute quantity, has a resonant fluorescence that is broad-band throughout the visible. However if one uses a tunable laser, especially in the near infrared, one can move into the aqueous window at 1.3 μm and use a sensitive detector in that region. One then finds that one can suppress all of that fluorescence and can have a very strong Raman frequency. The other point in support of the intense infrared tunable laser is that infrared and Raman spectroscopy are complimentary. In an aqueous solution, even though water absorbs very strongly in the infrared, the major advantage of an intense infrared laser source is that it can be focused on a small sample. By this means one can study micro-samples. More importantly, the very high intensity or spectral brightness will allow one to measure absorption in aqueous solutions, because, while the water is strongly absorbing, it is not totally opaque. A case in point was some recent work in Germany on measuring radicals in aqueous solution for a study of water pollution. That work is only possible with the spectral brightness of an infrared laser. So that, I think, is where we go from here.

Sir Harold Thompson: I agree, of course, with what has been said about the cleaning up of fluorescence in Raman work.

A discussion on the use of short pulses to remove line broadening then took place, initiated by Professor O. Svelto.

Professor O. Svelto: Someone commented before, saying that in biological substances, in particular in emission from liquids, we don't need frequency resolution because the lines are so broad. First of all, we should ask why the lines are broad. Very often they are broad due to inhomogeneous reasons arising from effects of the surrounding molecules. In a gas there is the possibility of removing the inhomogeneous broadening which is due to the mobile molecules. It is possible, in principle, in ordinary liquids, to get rid of the inhomogeneous effect in the same way. We have done some spectroscopy in the very-short time domain so that the molecules do not have time to move. In this way one can get rid of the inhomogeneous contribution and then the line is much narrower, even in a liquid state. In this case one can get much more information out of liquid states and also from solutions.

The question of frequency spread due to the shortening of the pulse was raised by Dr. E. R. Pike. Professor Svelto showed this to be small for a picosecond pulse compared with 200 Å. He suggested that one could narrow the line width to 100 Å which would be advantageous for research in chemistry and biology.

Dr. M. J. Colles pointed out that at any rate some of the inhomogeneous broadening is due to intermolecular interactions and has a spatial character as well as a time character. However, Professor Jacquinot thought that if one had a narrow enough linewidth in the laser one could pick selected sites, as one picks selected velocities.

Professor D. J. Bradley made an important point regarding the use of high-intensity short-pulse sources. One finds that the relaxation time in some instances depends on the intensity. It is therefore important to use as low an intensity as possible and as much intensification after it as available.

Dr. E. R. Pike: In my talk I explained how when one runs out of resolution with ordinary Newton (1760) prisms one has to go to electronic measures of amplitude fluctuation. Professor Bradley says he can do the same in measurement of short pulses, but I wonder why he wanted to do it. In other words, here one has a situation where conventional spectroscopy is most useful; and as one gets shorter and shorter pulses, then the prism becomes more and more useful, determining the pulse length from frequency spread.

Professor Bradley: The point I made was that one could extend the electronic technique into the picosecond regime. Basically, the interest in picosecond pulses is that one is able to look at molecules and solutions on a time scale on which they are standing still. One can do experiments with molecules in excited states on the time scale that vibrations relax. More importantly, for instance in the laser compression experiment, in which one might want to compress matter to a density of 1000 gcm^{-3} and a temperature of 10^8K one can only keep it together by inertial confinement for about 10 psec. We started off by using the frequency spread, but it is only now that we have learned to measure pulses properly. We can now produce pulses which obey the uncertainty relationship between duration and bandwidth as Professor Kaiser indicated. It is only in this special case that pulse duration can be inferred from a simple spectral measurement.

Dr. M. J. Coles: I would like to ask Professor Bradley if he can make some comment about another technique for short pulse measurement, which I think has proved to be very successful. That is the scheme proposed by Duguay of using an ultra-short pulse optical Kerr-effect shutter, which has resolution of the order of 1 psec. It has also been tied in with experiments by Rentzepis with echelons and time resolutions of the same kind of order have been achieved. I think that the advantage of this, as Professor Kaiser has indicated, is that a cell of CS_2 is not costly and the streak camera costs about £20,000! There may be some very specific reasons why one should prefer a streak camera for time durations greater than 1 psec. For shorter durations the reasons are more obvious.

Professor Bradley: I think the basic difference is that the Duguay shutter gives a nonlinear measurement. What one is effectively doing is using a mode-locked laser pulse to open the shutter, so really performing a convolution of the trigger pulse with the event to be measured. Thus it is not a direct linear method. The advantage of the direct photo-electric method is that the number of photons required to trigger the camera is very small indeed, about 100 photons. That is because of the image intensification in the system. Very little current is drawn from the photocathode. Since one is detecting single photo-electrons, this means that the camera technique is applicable to incoherent light, for looking, for example, at fluorescence. The Duguay shutter gives a time measurement of a pulse, particularly if one knows it is to be a good pulse, but it is not a direct linear measurement. The real difference is that, if one has a direct method of measuring, gradually one gives up using nonlinear techniques. If one wishes to look at very weak light sources one is not then tied to a laser at all. Streak cameras can give picosecond time-resolution over a wide range of wavelengths from X-rays to the near infrared while the Kerr-effect shutter is limited by the transmission properties of the cell.

Sir George Porter: I wonder if I can ask Mr. Euan Reid to say a word because he has done a number of measurements with a Duguay shutter; and then he did those measurements using Professor Bradley's camera. So he is in a fair position to comment.

Mr. E. S. Reid: The Duguay shutter can be used quite effectively for measuring spontaneous fluorescence. Unfortunately, the sensitivity is poor and the actual time involved is considerable. The measurements can be made much more rapidly with the streak camera.

Sir George Porter: One of the basic problems is that, to get the whole record with a Duguay shutter, one has to do a number of shots. If one could get a thoroughly reproducible picosecond pulse, then one would be in much better shape. The streak camera just needs one pulse to get all the information.

Dr. M. J. Colles: I don't think that is strictly true. In Duguay's experiment when he builds essentially a picosecond-sampling oscilloscope, it is on a single pulse basis and he can do time measurements with a single pulse in a few picoseconds. I agree that it is a nonlinear process and therefore, in that sense, the direct measuring technique is to be preferred. But I don't want people here who don't know about ultra-short pulse measurements to get the impression that it isn't possible to do simple experiments. If one has a very specific type of experiment of the type of Rentzepis, one can apply the shutter. I agree it has its limitations.

Line Narrowing by Time-biasing in Fluorescence

by G. W. Series, University of Reading, J. J. Thomson Physical Laboratory, Whiteknights, Reading RG6 2AF, England

Delegates may be interested to hear of a technique which reduces linewidth below the limit imposed by the natural width of energy levels. The technique is a development in the field of "coherence spectroscopy" referred to yesterday by Professor Jacquinot.

Suppose we are interested in structure (e.g. hyperfine structure) in the excited levels of an atom or molecule. Figure 1 illustrates, for example, two levels, each of natural width Γ, separated by the interval ω_0. It is often possible to excite the levels coherently by a pulse, either of light or of electrons or other charged particles or by passing a beam of ions through a foil

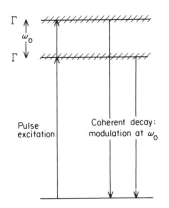

FIGURE 1. Pulse excitation followed by modulated fluorescence.

(beam foil spectroscopy). The conditions for coherent excitation are that the spectrum of the pulse must be wide enough to embrace the levels concerned, i.e., that the duration of the pulse be small compared with the period $2\pi/\omega_0$. Under these conditions the fluorescent light from the two levels is coherent and the intensity as a function of time shows modulation with period $2\pi/\omega_0$ under an envelope representing damping with decay constant Γ (Fig. 2). The Fourier transform of the decay curve yields the lorentzian (Fig. 3), peaked at ω_0 and of width 2Γ at half intensity. This is the spectrum of the intensity modulation, not of the amplitude of the fluorescent light. If there are more than two levels the spectrum will contain correspondingly more components. Whether or not these components can be resolved will depend on their separation in relation to the linewidth. It may be an advantage to be able to reduce the linewidth. This is accomplished by the technique we propose to describe.

Suppose that, in taking the Fourier transform of $I(t)$, we introduce a function $f(t)$ which will weight later times more heavily than earlier times. Then, because we are biasing in favour of the longer-lived atoms in the

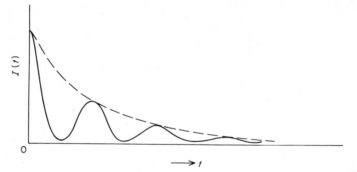

FIGURE 2. Modulated exponential decay of intensity.

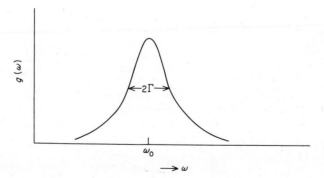

FIGURE 3. Fourier transform of modulated intensity.

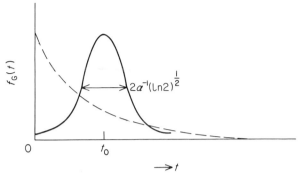

FIGURE 4. Gaussian biasing function, $\exp - [\alpha^2(t - t_0)^2]$.

assembly, the linewidth will be reduced below that obtained from an unbiased sample. The profile $g(\omega)$ of the spectral lines will depend on the particular form of the function $f(t)$:

$$g(\omega) = \int_0^\infty I(t) \cos \omega t f(t)\, dt. \tag{1}$$

If $f(t)$ is the step function:

$$f_s(t) = 0, \quad t < t_0; \quad f_s(t) = 1, \quad t > t_0, \tag{2}$$

then the profile of Fig. 3 becomes narrower around ω_0, as expected, but oscillatory structure develops in the wings. Such structure may introduce undesirable complication into a spectrum. But by inversion of equation (1) we may obtain an expression for $f(t)$ which will give any desired profile. We have found it useful to employ the gaussian function (Fig. 4):

$$f_G(t) = \exp - [\alpha^2(t - t_0)^2]. \tag{3}$$

If the width parameter α is chosen in relation to the position t_0 of the peak according to the relation:

$$\alpha^2 = \tfrac{1}{2}\Gamma/t_0, \tag{4}$$

then the lorentzian of Fig. 3 is converted into the gaussian $\exp - [(\omega - \omega_0)/2\alpha]^2$. The oscillatory structure has now been eliminated. The half-width $4\alpha(\ln 2)^{\frac{1}{2}}$ can be controlled by choice of t_0 or α. (Fig. 5)

We pay a price for this reduction in width: insofar as we have thrown away the earlier part of the signal, the information we are using is exponentially attenuated and the signal-to-noise ratio of the spectrum deteriorates. For this reason I do not think that the technique offers advantages in locating the peak of an isolated line. I *do* think it offers advantages in locating the peaks

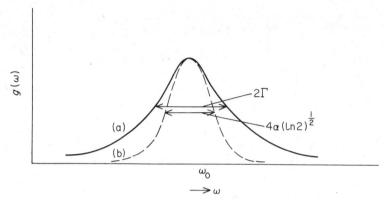

FIGURE 5. Narrowing achieved by biasing.

of partially-overlapping or unresolved structures. The usual approach to this problem is by a de-convolution technique: such techniques rely heavily on knowledge of the line profile. If the line profile is not known with confidence, (departures from the lorentzian have been found in practice) then the technique of narrowing by time-biasing gives the possibility of improved resolution, with reliance on the line profile only for the calculation of small correction terms. The well-known property of the gaussian, that its wings fall much more rapidly than do those of the lorentzian, is particularly advantageous in improving the resolution of overlapping lines.

The appended references give examples of modulated fluorescence following pulse excitation and of the application of the line-narrowing technique in a closely related field, "level-crossing spectroscopy". The technique has not yet, so far as we are aware, been applied in the way described.

REFERENCES

Modulated fluorescence
Alexandrov, E. B. (1964), *Optika Spektrosk.*, **17**, 957.
Dodd, J. N., Kaul, R. D., and Warrington, D. (1964), *Proc. Phys. Soc.* **84**, 176.
Hadeishi, T., and Nierenberg, W. A. (1965), *Phys. Rev. Lett.*, **14**, 891.
Haroche, S., Paisner, J. A., and Schawlow, A. L. (1973), *Phys. Rev. Lett.*, **30**, 948.

Time-biasing
Deech, J. S., Hannaford, P., and Series, G. W. (1974), *J. Phys. B: Atom. Molec. Phys.*, **7**, 1131.
Figger, H., and Walther, H. (1974), *Z. Phys.*, **267**, 1.

Where do we go from here?

PROFESSOR SIR HAROLD THOMPSON:

This symposium has been a great success. A very high level of thinking and talking has been maintained throughout. A high level has also been maintained in the discussion.

There has been astonishing progress in these fields in recent years. Twenty-five or thirty years ago I used to go down to Malvern to fetch for my own use in the University an occasional photoconductive cell—lead telluride or whatever it might be—to put into a grating spectrometer and measure the rotational fine structure of vibration bands of polyatomic molecules. When I went into the infrared business, a resolving power of 1 cm^{-1} or 30,000 MHz, was considered to be pretty good. By the end of the war, with the photoconductive detectors, we began to get 0.1 cm^{-1} or 3,000 MHz, while today we see that one can now get 100 MHz or even less!

We used the isotope splitting of hydrogen bromide—79 and 81 isotopes—as a routine check on the performance of our best grating spectrometers, and the resolution wasn't complete; it was just that one did see two peaks half-separated. If we did that, we thought we were doing pretty well. We saw yesterday, these things spread yards apart. It is rather like the transition in the text books of physical chemistry and physics, where one sees the isotope splitting in the vibration rotation bands of hydrogen chloride—the chlorine isotope split—shown over the years. Yesterday, with HBr and HCl one has it split still wider apart, showing that there has been a fantastic increase in resolving power and achievement.

In this meeting the subjects discussed have covered a very wide range indeed. There were not only very striking developments in technique, but applications and implications for physics, chemistry, biology and astronomy. Of course a great deal of it was concerned with tunable lasers and pulse techniques used in one way or another. Professor P. Jacquinot gave us a splendid review of new ways of getting very high resolution, and illustrated them by his examples taken from hyperfine structure of atomic levels.

Professor S. D. Smith described the spin-flip method. He illustrated its use in studying isotope splitting in the sub-structure of nitric oxide lines, in bands of carbonoxysulphide and other molecular spectra. He showed us how this method can be developed for cw spectroscopy.

The new wide coverage of a spectrum using tunable lasers is impressive and could of course be very important for many applications.

Dr. E. R. Pike described to us photon correlation spectroscopy with some fascinating applications, thermal diffusion in liquids, measurement of the size—a fascinating thing for a physical chemist—of a hydration shell round an ion in solution. It was used also to study vacancies in solids, in plastics, and to measure some biologically interesting molecules which may have very practical importance. I remember, too, the fluid-flow problem in blood behind a retina and in wind tunnels. These things cover an enormous range of applicability.

Dr. A. Mooradian gave us further examples of the wide wavelength coverage now obtainable with tuned lasers and mixing lasers, using some of those semiconducting materials. He suggested some possible high-energy fixed-frequency sources which might be of considerable importance in photochemical work.

Professor D. J. Bradley and Professor W. Kaiser gave us excellent accounts of work with ultra-short pulses using various kinds of laser, and suggested uses in photochemistry, even for such fantastically different things as continental drift, the predetection, possibly, of earthquake phenomena as well as more routine applications in studying molecular vibrations and some aspects of molecular dynamics.

Dr. R. G. Brewer described the Stark switching method and some possible applications using it, including, possibly, studies of molecular energy relaxation.

The nonlinear optical techniques which Professor S. E. Harris described were very interesting for me and also, I believe, for Professor W. C. Price, especially the possibility of using them to produce vacuum ultraviolet sources.

Professor C. H. Townes gave us a fine account of developments in astronomy over a very wide spectral range, and showed us spectral characteristics discovered in stars and interstellar space including bands of simple molecules such as carbon monoxide, carbon dioxide, ammonia and other things. It was intriguing to notice this apparent variation of the predominance of isotopes in different stars, in different circumstances. He described to us some very important technical developments. The conversion of infrared into visible radiation and the new bi-telescope interferometric device. These methods seem to have very great potential power.

Professor W. C. Price gave us a survey of some of the principles and results of recent work in photoelectron spectroscopy using ultraviolet photons rather than X-rays, and the determination of molecular orbital energies equivalent to ionization potentials from levels and their correlation in series of molecules. I think that here is a need for new sources of radiation— monochromatic sources—in the region between the current helium-1 or helium-2 lines and the conventional X-ray lines used in this work. Perhaps some of the kinds of things that have been discussed at this meeting may lead to developments of such sources at higher energy in the middle region.

As a chemist, the thing which intrigues me with photoelectron spectroscopy is the correlation, if it is possible, of these ionization potentials of molecules with what we know about the chemical reactivity of molecules and their tendency to form complexes with other molecules. To some extent this work has given information of that kind. We didn't hear from Professor W. C. Price much about the results with X-ray photoelectron spectroscopy which, if properly applied and developed, has probably more practical use in industry and elsewhere than ultraviolet photoelectron spectroscopy. This latter is very important to us in determining molecular orbitals, but it is limited in scope because for the most part it concerns measurements in the vapour state and many substances cannot be put satisfactorily at the right vapour pressure into the vapour state. Although nowadays we are working at somewhat higher temperatures, there are still some problems of a practical kind in making the measurements.

We had a very enlightening talk from Professor Britton Chance about these highly complex biochemical matters, electron transfer processes in the living cell. He made very elegant use of ultraviolet absorption, laser pulses and other physical methods, and made his measurements within very short time intervals. He gave us what he called a "Cook's Tour" of the formation of ATP and discussed many aspects of photosynthesis.

Then we had an interesting talk from Mr. E. J. Millett about the use of computers in increasing the power of spectroscopy as a research tool, and as a means of industrial control.

So much for my view of the factual side of this meeting. In these days when the practical relevance of science is being questioned in high places, we may be asked what there is here for the industry and for applied science. Will there be anything, for example, like the routine application in industry of infrared spectroscopy? Indeed I realize there are other techniques such as mass spectrometry and spin resonance which may have now superseded infrared work for many analytical purposes. It is extremely difficult to answer such questions at present, and certainly I would not wish to have it said that I was, in the slightest, of the opinion that pure research should not be encouraged. We have to look at this factually. It may be that, arising from some of the work which we have discussed in the last two days, sharp monochromatic sources will be developed for use in some simple instruments for chemical analysis. If that is so, it will be very useful, provided the instruments are reasonably cheap.

As you know, for many years past it has been the hope that by having very intense fairly monochromatic sources one should be able to initiate certain chemical reactions which would lead to chemical products of great value. I think this still stands. I don't know why it hasn't worked so far but I don't think it's entirely because one has not available a whole range of monochromatic sources; there are deeper issues involved. This would be an admirable way in which we could service industry, if one converted the chemical industry from a catalysis industry into a photochemical industry.

Secondly, when asked these questions by governments and the people who have money, we must always admit to them that one should never use expensive equipment and complicated instruments in industry when cheaper ones and simpler ones can adequately do the same job. I am bound to say that some of the very recherché instruments that have been described here, are still not in the state or range suitable or even required in industrial work.

My third point is that we have still to consider whether higher resolving power is really going to lead to more information. Without a doubt a lot of the things described do lead to much more information. But I am talking at this moment of practical things where it may be that, having much higher resolving power would cost much more and would not lead to much more of the sort of information that industry wants. I emphasize this because we are fighting to acquire money for scientific research these days, and we have to satisfy the people who have the money.

We must now further pursue the question "Where do we go from here?" One thing is absolutely certain—that the technical methods which have been described yesterday and today will go on being developed further. It is probably also true that materials for new lasers will be discovered, and some of them will be especially useful for shorter wavelengths. There seem to be possibilities for the development of new vacuum ultraviolet and X-ray sources which could have uses in other techniques. One uses the existing techniques to make measurements in the search for more detailed natural knowledge, and then that will lead in its turn to the stimulation of the need for better instruments which will then be developed, and the cycle will go on.

It is also important to say that the discovery of things of practical utility is often completely unforeseeable, unforeseen and accidental. Even so, if one can strike them occasionally it is very important. I often quote one case of that kind which I am sure most of you know about. Infrared spectrometry used today for fuel analysis, chemical analysis, studying the structure of synthetic polymers and all sorts of things of this kind really was the result of the discovery of the isotope of hydrogen. The sequence of thinking was simply this: hydrogen in 1932 had two values for the atomic mass which differed—that from mass spectrometry and that from chemical reactions. Scientists asked why, and this led to the suggestion that there was an isotope. Then they separated it and the isotope deuterium was discovered. A few years after that, H. S. Taylor of Princeton decided to study the thermal reaction between deuterium and methane. Of course he got an ungodly mixture of isotopic species, monodeuteryl, dideuteryl, trideuteryl methane, with no simple method of analysis of the mixture. That is how vibrational spectroscopy really got afloat, and this is now the basis of all large industrial applications of the infrared. Sir Gordon Sutherland and I immediately went to British instrument makers and begged them to make some infrared instruments with complete lack of success!

I have one final comment about the future. The immediate future may be determined in large measure, perhaps considerably, by the availability or the lack of money.

PROFESSOR D. J. BRADLEY

I think that rather than using laser techniques, simply as a new way of monitoring what is already known in industry, the real future must lie in using lasers to produce new materials. Where I see the hope for the future is in the fact that lasers for the first time are becoming efficient enough to be thought of as energy sources. For instance, using E-beams it is now possible in the vacuum ultraviolet to get high-powered tunable lasers operating around 1700 Å. As Professor S. E. Harris made it clear, with this approach, lasers could be developed over a very wide range of wavelengths. Perhaps what will happen is that we shall start off using these tunable lasers, for instance, to manufacture fine drugs. This is one of the questions I should like chemists to think about. If we could supply an efficient source, narrow-band or broad band, but tunable throughout the ultraviolet, visible and infrared, could one think of using this in chemical processing, not just to monitor the processes which are already operating, but to create new materials by techniques similar to photosynthesis? This is perhaps, too, a possible application for picosecond pulses, which might even make a more important contribution in this way than in application to laser fusion.

PROFESSOR C. H. TOWNES

I was very much impressed by the general philosophical view towards applications in science and their interaction, given to us by Sir Harold Thompson. In this particular example of deuterium the key role of infrared spectroscopy was new to me and quite fascinating. Looking at the situation generally with respect to what we are doing here, or have done during the last two days, it seems to me that a great deal of what we might call pure science is going to be something for pure scientists. In the course of doing this kind of work, ideas, techniques, new discoveries somewhat by chance are going to be developed which can be enormously important. The laser field itself, one can reasonably say, came about as a result of the study of interaction between microwaves and molecules—a pretty esoteric subject, but which in turn came about as a result of applied work in radar. This interaction is very characteristic.

If one looks at the laser field in general we can be very encouraged about its potentiality in terms of giving more power to the human race. Whether this is good or bad is a more detailed, broader question. Consider the general importance of light, optics, electromagnetic waves to mankind, the general power of this electronic control which we are now acquiring over light and electromagnetic waves of shorter wavelength than we had before, there is scarcely a field in human activity that is not going to be substantially touched by it. The question is not "can one do almost anything with it?" but "in what way is it going to be competitive with other techniques?" One can already see in some surprising areas how quickly it is growing. In civil engineering, for example, most of the really respectable firms in the United States use lasers, simply for drawing straight lines, sometimes in measuring distances. Lasers are beginning to be introduced into supermarkets to read the numbers on packages quickly and efficiently, and they can do a variety of other things. I give these examples simply because they are mundane things so far as scientists are concerned, and perhaps we don't pay very much attention to them, yet they are areas that illustrate the general breadth of the technique.

The contribution we can particularly make in this direction is not to abandon basic science, but rather to be sensitive to the general needs of society. We should also be somewhat acquainted with applications and needs of industry so that we are ready to make suggestions and transfer information and watch out for those things which can be useful. This is not, perhaps, confined to basic science, but it may be *our* best role: to be sensitive to those needs, aware of them, conscious of them, and see that they get fulfilled.

Finally, I should like to say a few words about the future of astronomy. The first point is that within the last decade there have been a number of startling basic discoveries in astronomy. There is every reason to believe that this is going to continue. Secondly, of the range of things discussed here, within about a decade we have the possibility of solving most of those technical problems—instrumental problems—and being able to bring this large range of frequencies into the kind of usability, and within the reach of ultimate limits, that we find in optical and radio work today. What new discoveries will come from it? One can be confident that there will be many and exciting ones, but there is another realm of astronomy which is certainly going to be a part of the future and which I neglected completely in my talk.

I mentioned first that in astronomy one has the problem of taking what radiation is there and doing the best we can with it—one cannot play with the system. That is not entirely true. Within our solar system we can, as in the radio region, do active work which will probably be quite crucial in establishing theories of gravitation, in particular, with very accurate measurements. The accuracy surely is going to be somewhat connected with the wavelengths involved. We should be moving to shorter wavelengths in order to do that well.

Rather more generally I would like to bring up the future of space astronomy. Our atmosphere gives us two kinds of limitations: one is imperfect seeing and that means more than just a blurry image: it also costs us sensitivity. When we get above the atmosphere for astronomy there, the concentration of light into smaller regions, which is possible as a result of more perfect seeing, essentially means a gain in signal-to-noise ratio which will allow us to see further back in time and be able to look at more obscure objects by roughly a factor of 100 in the visible region. That's quite a lot! Furthermore in the infrared region there is another very important factor which space work will take care of for us. The limit against which we are now trying to work basically is thermal radiation in the infrared region. That thermal radiation is an inherent property of our atmosphere. If we can get above the atmosphere we are down to temperatures of the order of 3 K, crudely speaking, instead of 300 K. In considering the exponential in the Boltzmann factor—that means an enormous factor in the background radiation against which we would have to work. We can work with cooled instruments, and this, too, will benefit us enormously not just in the transparency of the atmosphere but also in the signal-to-noise. So with the availability of telescopes above the atmosphere, particularly with enough weight capacity so that one can be flexible in the instrumentation used, I think there will be a very powerful new kind of astronomy that will continue to develop.

PROFESSOR BRITTON CHANCE

I don't think those concerned with inner space can be as bold in their predictions as for outer space. Perhaps we won't find out so much new but we may be able to make it available to more people and to the medical sciences, and therefore promote health and welfare which is one of the goals we are thinking about. While this conference was identified with very high resolution spectroscopy, it is true that many diseases are related to the very low resolution spectroscopy. They are diseases of haemoglobin—for example, sickle-cell disease. There is not a quick and ready way of identifying in a blood sample, for example, which has the sickle trait, which has the altered residue and which doesn't. If Laser Raman spectroscopy were to provide an indicator of the residue that would be fine; but this is a very simple low-resolution thing. Laser Raman spectroscopy has not yet shown that it can do this. As I mentioned briefly, Spiro has difficulty even using Raman spectroscopy as an indicator of the spin-state of the ion, which we would think he ought to be able to do. Probably he will be but it's at present highly controversial.

The second general thing is that optics has the great advantage of being non-destructive to the general world, but of course we realize its dangers, especially when we use very short high-intensity pulses. But in steady state phases it is one of the accepted non-destructive approaches, and indeed there is nothing the clinician likes better to do than to probe the openings of the body. We haven't used this with light-guide techniques in a sophisticated way, whereby the signals from haemoglobin which usually obscure whatever one sees optically, are got out by simple computer techniques. One knows how much haemoglobin is there, and by simple "computery" one can get rid of haemoglobin signals and look for the more interesting signs and symbols which would be characteristic of the state of health and disease of the organ. Of course one of the diseases of the organ which I am sure is affecting us right now is that oxygen doesn't get delivered to our tissues and so our brain cells die off at an even higher rate than the one billion per day or whatever rate it is at which they are supposed to die off. So I think that, just in stressing the point, there are many applications where the manufacturers, or perhaps the busy clinician, just don't get to and we have somehow to bridge the gap.

Of course on the scientific side there are myriads of applications where better signal-to-noise ratio and even better resolution would help.

Concluding Remarks

DR. F. E. JONES

I think Lord Rank would have been pleased with the way that we have begun to spend the money furthering the sciences he had faith in—opto-electronics and nutrition. One thing I am certain about, and that is the privilege and pleasure of initiating and organizing this conference—speaking for myself and my colleagues on the Committee and Mr. Hadley—there is no doubt we are one hundred per cent rewarded by the response during these last two days. It really has been overwhelming for all of us and has generated a quite warm glow.

I said earlier that I thought that after discussing this for years and writing to you all, that it was rather like lighting a bonfire which burned for two days and then went out. It's just not true, of course. The whole thing was self-combustible right from the time Professor Jacquinot started. Unfortunately we now have to damp it out—it's not going to burn out, so we shall damp it out. So may I thank you all for the response, and for coming and joining in this two-day session, and I wish all of you who have to travel far a very speedy and safe journey back home.

INDEX

A

acoustic phonons, 61
alkali metal vapours, frequency tripling in,
   cadmium, 144, 148
   rubidium, 144–147
anti-Stokes scattering, 120–125
argon ion laser, 106
astronomy, 159–185
   atmospheric transmission, 163–167
   sky noise, 167
   spatial interferometry in, 178–184
   spatial resolution in, 173–174, 176–184
astronomical spectra,
   carbon dioxide in Martian atmosphere, 176, 177
   carbon monoxide in Orion, 162–164
   fine structure of ionic species, 172–173
   hyperfine structure of ammonia, 161
   R. Leonis, 159–160
   Mercury, 182–184
   Moon, 160
   nebular water maser, 160–161
   stellar atmospheres, 169–171
atmospheric transmission, 163–167
atomic spectroscopy,
   absorption line narrowing, 3, 5, 8–10
   atomic beam techniques, 3–5
   Fabry-Perot spectroscopy, 3, 6, 11
   fluorescence line narrowing, 3, 5–6
   fluorescence line narrowing by time biasing, 257–260
   Fourier transform spectroscopy, 1–3, 11
   Quantum beats, 10–11
   saturated absorption spectroscopy, 3, 5–6, 245
   two-photon excitation, 5–8, 246, 248, 249
   two-step excitation, 5–8

B

Brownian motion, 61
Brillouin scattering, 61, 62, 65
biological spectroscopy (see also cell spectroscopy),
   cell spectroscopy, 205–226
   Raman spectroscopy in, 252–254
   scattering from biological samples, 66–70
   temporal resolution in, 251
   ultrashort pulse methods for, 215–226, 250–251
bonds, 191–202
bonding electrons, 189, 191, 194, 198–202
British Gas Board, 72

C

carbon dioxide laser,
   in the Stark-switching technique, 129
   pump for spin-flip Raman laser, 22, 23, 28, 29, 31, 33, 82, 89
   second harmonic generation of, 23–24
carbon monoxide laser, 83, 85, 86, 88
Carr-Purcell echoes,
   in $^{13}CH_3F$, 139–140
cell spectroscopy, 205–226
   carotenoid response, 212–226
   cell structure, 206–211

INDEX

photosynthesis, 211–226
  ultrashort pulse methods for, 215–226
coherence, 127, 244–246
coherent optical spectroscopy, 10–12, 127–142
  Carr-Purcell echoes, 139–140
  free induction decay, 128–134
  modulation resonance spectroscopy, 10–11
  optical nutation, 130–132, 247
  photon echoes, 129
  quantum beats, 10–11
  Stark-switching techniques, 128–142
computer systems, 234–242
  camac system, 238
  dedicated mini computer, 235–236
  MRL hierarchical system, 238–242
  multiple access computers, 236–238
  varian IBM "spectroshare" system, 237
computer techniques, 227–242
  analysis of complex line spectra, 230–232
  analysis of simple line spectra, 229–230
  band spectra, 234
  calculation times, 228–229
  Fourier transform spectroscopy, 233
  gas chromatography, 232–233
  Hadamard transform spectroscopy, 233
  multichannel spectroscopy, 233
correlation spectroscopy (see photon correlation spectroscopy)
critical opalescence, 61

D

de Haas – Van Alphen effect, 21
difference frequency generation,
  in $CdSeAs_2$, 88–89
  in InSb, 31–33
  in $LiNbO_3$, 87–88
  phase-matching, 33
  spectrometer, 88
difference frequency spectrometer, 88
doppler anemometer, 70
doppler broadening, 2, 4–6, 39–40, 248–249
  suppression in two photon transitions, 248–249
  in NO, 39–40
doppler limited spectroscopy, 39, 82
  of caesium, 82
doppler linewidth, 6
  in free induction decay, 133
doppler velocimetry, 67–73
dyes (see also dye laser),
  7-diethylamino-4-methyl coumarin, 93
  3,3′-diethyl oxadicarbocyanine iodide, 102
  1,3′-diethyl-4,2′-quinolyoxacarbocyanine iodide, 105
  rhodamine b, 93
  rhodamine 6g, 2, 93
  saturable absorber dye, 92, 94
dye laser, 15–17
  resolution, 17
  saturable absorption, 95, 101
  saturable amplification, 101
  spectral brightness, 17
  tuning by Fabry-Perot interferometer, 95, 101
  CW dye laser, 7, 87
    line width, 1
    modelocking, 105–107
  flashlamp pumped laser,
    amplifiers, 95–96
    modelocking, 94–96, 100, 101, 105

E

electron-optical streak camera, 97–110
  (see also streak camera)
entropy, 244–246
  Fokker-Planck equation, 245

F

Fabry-Perot interferometer, 1–3, 6, 11, 53, 61
  for dye laser tuning, 95, 101
Fabry-Perot spectrometer, 167–171
Faraday rotation,
  in InSb, 21
fine structure of ionic species, 172–173
fluid turbulence, 61, 71–73

272

fluorescence line narrowing, 3, 5–6
  by time biasing, 257–260
Fokker-Planck equation, 245
Fourier transform spectroscopy, 1–3, 11, 159, 167, 182–183, 128, 134 (see also optical pulse Fourier transform spectroscopy)
  computer techniques, 233
  noise, 167
Franck-Condon principle, 188
free induction decay, 128–134
  Doppler width, 133
  in NH_2D, 132
  interference pulses in $^{13}CH_3F$, 133–135
  Maxwell-Bloch equations, 133
frequency multiplication, 143–157 (see also third harmonic, second harmonic and difference frequency generation)
  by nonlinear Van der Waals interaction, 153–156

H

Hadamard transform spectroscopy, 233
Hanbury-Brown and Twiss
  interferometric technique, 176
heat pipe oven, 149
heterodyne techniques, 58–60, 160, 162, 163, 174–176, 178, 246–247
  noise, 175–176, 246–247
  resolution, 175–176
holography, 151–153
homodyne detection, 58, 60, 69
hydrogen fluoride laser, 82–83
hyperfine structure,
  of ammonia, 161
  of Sodium D lines, 4–5

I

inert gases,
  frequency tripling in,
    argon, 148
    helium, 145
    xenon, 148
interstellar cloud composition, 249 (see also astronomy)
ionization potential, 188, 198
isotope separation by tunable lasers, 5

J

Jahn-Teller splitting, 192–193

K

krypton-ion laser, 173

L

Lamb dip spectroscopy,
  in water vapour, 40, 86
lasers, (see also mode-locked laser and particular types of lasers)
  coherance, 127, 244–246
  phase stability, 58–59
laser radar, 61
line narrowing by time-biasing, 257–260
liquid crystals, 61

M

McMath solar telescope, 180–181
maser amplifiers, 160
Michelson interferometer, 2, 12
Michelson's stellar interferometer, 176, 178
mode-locked laser (see also individual classes of lasers),
  biological applications, 215–226, 250–251
  controlled fusion, 103, 255
  fluorescence lifetime measurements, 102
  interaction studies with, 111–125
  laser compression, 103, 255
  pulse measurement of, 91–110, 255–256
  self phase modulation, 103
  transform limited pulses, 96
molecular collision mechanism, 127, 134–142

molecular gas spectroscopy,
  pulsed spin-flip Raman laser
    spectroscopy, 30–34
    of deuterium bromide, 30
    of NO, 30
    signal-to-noise ratio, 31
  CW spin-flip Raman laser
    spectroscopy, 36–40
    carbonyl sulphide, 36–37
    Lamb dip in water vapour, 40
    stibine, 37
molecular radii, 65
natural linewidth, 40

N

neodymium:glass laser,
  frequency tripling, 144–150
  mode-locked, 91, 97, 106, 108, 215
neodymium:YAG laser,
  in nonlinear optics, 87
  mode-locked, 144, 147

O

Olympus 593 (Concorde) engine, 72–73
optical Kerr shutter, 255–256
optical nutation, 130–132, 247
  in $^{13}CH_3F$, 130–131
  Maxwell-Schrödinger equations, 130
optical parametric generator, 114–116
  conversion efficiency, 115–116
  tuning for a $LiNbO_3$ crystal, 115–116
optical parametric oscillator, 14–17, 89, 160
  resolution, 17
  spectral brightness, 16
optical phonons, 61, 125
optical pulse Fourier transform
    spectroscopy, 128, 134 (see also
    Fourier transform spectroscopy)
opto-acoustic spectroscopy, 41–44
  of carbonyl sulphide, 41, 43–45
  of NO, 42
  resolution of, 44

P

parametric amplifiers, 160
parametric interactions, 112, 113 (see
    also optical parametric generator
    and oscillator)
perturbed atomic fluorescence
    spectroscopy, 12
photon correlation spectroscopy, 51–73
  biological applications, 66–70
  detector response time, 52, 54
  first-order correlation function, 52
  fluid turbulence, 71–73
  heterodyne techniques, 58–60
  homodyne techniques, 58, 60, 69
  intensity correlation function, 57
  liquid diffusion, 63–64
  molecular radii, 65
  photon correlation function, 60
  photon fluctuations, 56–60
  second-order correlation function, 60
photon echoes, 129, 134–142
  Carr-Purcell echoes in $^{13}CH_3F$, 139–140
  dephasing time, 135, 139
  three-pulse photon echo in $^{13}CH_3F$, 138, 146
  two-pulse photon echo in $^{13}CH^3F$, 137, 140–141
plasmons, 61
photoelectron spectroscopy, 187–203
  halogen derivations of methane, 194
  hydrides isoelectronic with inert gases, 189–190
  ionic molecules, 195–198
  ionization potential, 188, 198
  multiple bonded diatomic molecules, 198–202
  photoionization, 187
  UV photoelectron spectroscopy, 187
photoionization, 187
pressure broadening,
  in carbonyl sulphide, 39
  in NO, 39
pulse selector, 112–113

Q

quantum beats, 10–11

INDEX

R

Raman scattering, (see spin-flip Raman scattering, spin-flip Raman laser, anti-Stokes scattering, spontaneous Raman scattering and stimulated Raman scattering)
Raman spectroscopy,
  in biology, 252–254
Rayleigh scattering, 61–62
resolving power, 2–3
Royal Aircraft Establishment, 71
ruby laser,
  in nonlinear optics, 87, 146
  mode-locked, 106

S

saturated absorption spectroscopy, 245
second harmonic generation,
  conversion efficiency, 24
  in tellurium, 23–24, 31
Seigert relation, 60
self-phase modulation, 103
semiconductor diode laser,
  grating controlled, 79–81
  laser medium,
    GaAs, 79–80
    $Pb_x Sn_{1-x} Te$, 77, 80
    $PbS_x Se_{1-x}$, 81
  pressure tuning, 81
  resolution, 17
  spectral brightness, 16
  spectral output, 75
sky noise, 167
spatial interferometer, 178–184
  noise, 184
  study of Mercury, 182–184
spatial resolution, 173–174, 176–184
  Hanbury-Brown and Twiss interferometer, 176
  Michelson stellar interferometer, 176, 178
  spatial interferometer, 178–184
spectral brightness, 13, 15–16
  dye laser, 16
  optical parametric oscillator, 16
  semiconductor diode laser, 16
  spin-flip Raman laser, 16

spin-flip Raman laser, 13–49, 82–87
  axial modes, 34–35
  Doppler limited spectroscopy, 39
  gain, 36
  InSb cavity, 33
  linewidth, 27, 30–31, 34–35, 84–85
  mode control, 46–47
  output power, 27–28
  pumping, 22–24, 28–29
  resolution, 17, 47
  spectral brightness, 16
  spectral output, 85
  spectral stability, 85
  spectrometer, 28–29
  spectroscopic applications, 30–31, 36–38, 41–44
  stabilization, 85
  threshold, 27
  transverse modes, 35
  tuning, 22–25
spin-flip Raman scattering,
  spontaneous spin-flip Raman scattering linewidth, 84
  stimulated spin-flip Raman scattering,
    in CdHgTe, 47
    in $Hg_x Cd_{1-x} Te$, 87
    in InAs, 86
    in InSb, 18, 21–22, 26–27
spin waves, 61
spontaneous Raman scattering, 18, 61
  linewidth, 121–124
Stark effect, 197
  in optical nutation, 247
Stark-switching technique, 127–142
stimulated Raman scattering, 112, 117–125
  dephasing time, 118–119, 121
  relaxation time measurements, 118–124
Stokes-Einstein relation, 65
streak camera, 97–110, 255–256
  calibration, 100
  deflection plates, 97
  extraction mesh, 98–100
  optical delay for, 99
  phosphor, 97
  photocathode, 97–100
  Photocron II, 104–105
  resolution of, 103–105
  spatial resolution, 103
  X-ray camera, 108–109

INDEX

T

temporal resolution,
  in biological spectroscopy, 251
  photodiode, 97
  photomultiplier, 97
  streak camera, 103–105
third harmonic generation, 143–157
  application to holography, 151–153
  by focussing, 146–147
  gas breakdown in, 148
  in alkali metal vapours, 144–148
  in inert gases, 148
  phase matching, 145
  of neodymium:glass laser, 144
  of neodymium:YAG laser, 144, 147
  of ruby laser, 146
two-photon transitions, 246
  in sodium, 6–7
  suppression of Doppler broadening, 248–249
two-step transitions,
  in sodium, 6–7

U

up-conversion in gallium phosphide, 249
up-conversion spectroscopy, 173–175
  hydrogen chloride, 174–175
UV photoelectron spectroscopy, 187

W

water absorption spectrum, 76–78
  Lamb dip spectroscopy of, 40, 86
  nebular water maser, 160–161

X

X-rays,
  from laser produced plasmas, 107–109
X-ray lasers, 103, 108
X-ray streak camera, 108–109